King Sequoia

King Sequoia

The Tree That Inspired a Nation, Created Our National Park System, and Changed the Way We Think about Nature

William C. Tweed

Heyday, Berkeley, California
Sierra College Press, Rocklin, California

This Sierra College Press book was published by Heyday and Sierra College.
© 2016 by William C. Tweed

All rights reserved. No portion of this work may be reproduced or transmitted in any form or by any means, electronic or mechanical, including photocopying and recording, or by any information storage or retrieval system, without permission in writing from Heyday.

Library of Congress Cataloging-in-Publication Data
Names: Tweed, William C., author.
Title: King sequoia : the tree that inspired a nation, created our national park system, and changed the way we think about nature / William C. Tweed.
Description: Berkeley, California : Heyday, [2016] | Includes bibliographical references and index.
Identifiers: LCCN 2015049651| ISBN 9781597143516 (pbk. : alk. paper) | ISBN 9781597143561 (e-pub)
Subjects: LCSH: Coast redwood–California–History. | Sequoia National Park (Calif.)–History.
Classification: LCC SD397.R3 T84 2016 | DDC 634.9/7580979486–dc23
LC record available at http://lccn.loc.gov/2015049651

Cover Art: Ranger Doug "Ranger of Lost Art" rangerdoug.com
Cover Design: Diane Lee
Cone Illustrations: Joe Medeiros
Interior Design/Typesetting: Leigh McLellan Design
Printed on demand by Lightning Source, USA

King Sequoia: The Tree That Inspired a Nation, Created Our National Park System, and Changed the Way We Think about Nature was published by Heyday and Sierra College.

Published by Heyday
P.O. Box 9145, Berkeley, California 94709
(510) 549-3564
heydaybooks.com

10 9 8 7 6 5 4 3

Do behold the King in his glory, King Sequoia!
—JOHN MUIR, 1870

Contents

Foreword by Joe Medeiros ... ix

Introduction ... xv

1 The Mammoth Trees of California ... 1
2 To Name Is to Know ... 21
3 A Grove Called Mariposa ... 29
4 An Arboreal Mecca ... 41
5 Yet Grander Forests ... 51
6 A Wandering Scot ... 59
7 Free for the Taking ... 75
8 Of Tunnel Trees and National Parks ... 91
9 For the Greater Good ... 107

10	A Source of Inspiration	123
11	Science and Time	137
12	Running into Limits	147
13	Words as Grand as Trees	165
14	Belonging to All	181
15	Kindled Light	199
16	Worth the Fight	213
	Epilogue	229
	Acknowledgments	233
	Notes	235
	Bibliography	249
	Index	259
	About the Author	265
	About the Sierra College Press	266

Foreword

Many years ago, when I was a young ranger-naturalist at Devils Postpile National Monument, my boss announced that "the brass" were coming for an inspection tour. Our tiny monument was managed under the distant Sequoia and Kings Canyon National Parks, and the visiting party included Ranger William Tweed. My supervisor told us to look sharp and to polish our badges and belts. As they still do today, our official National Park Service uniforms bore insignias with images that represented various aspects of the protected lands. The shoulder patch, shaped like an arrowhead to reflect the rich historical components of the service, featured the American bison and the sequoia—representing our animals and plants, respectively—both standing proudly in front of a majestic snow-clad mountain. Our belts and the bands of our flat-brimmed hats were stamped with sequoia cones and foliage. As a biologist and a Californian, this iconography resonated with me, although it wasn't until I read the book you hold in your hands that I understood why these giant trees, which grow naturally in only one state, graced the uniforms worn by all NPS employees across the nation.

At the time, my work was to interpret the geology of the spectacular volcanic columns of the Postpile, tucked high in the Sierra Nevada within a lodgepole pine and red fir forest. As a new ranger, I worried that Tweed, an interpretive specialist for the parks, would scrutinize my knowledge and teaching style. Thankfully, my worries were put to rest when this short visit by the Sequoia-Kings crew turned out to be filled with only grace and enthusiasm. More than three decades later, I still reflect on this experience as both enriching and enjoyable.

Now, after a long career of teaching college botany and ecology, my path once again crosses Tweed's, and I find myself thoroughly enjoying this opportunity to write a foreword to his newest book, which the Sierra College Press actively sought to copublish with the great folks at Heyday. This book represents the fifth such collaboration of Heyday and SCP, a partnership that has benefitted from the latter's focus on the Sierra Nevada and the former's stellar efforts of celebrating California as a whole, particularly its rich history, culture, and natural history. This latest copublication has produced a wonderful book that serves those interested not just in California and the Sierra Nevada specifically but also in the larger history of how modern humans interact with the natural world. This book is not so much a story of the sequoia itself as a story about what the trees have meant to people, and how this relationship has shaped the futures of both California and the United States. It is not coincidental that we timed the release of this book to acknowledge and celebrate the proud centennial of the establishment of the National Park Service.

Sequoias and their relatives flourished widely during the Mesozoic Era, more than seventy million years ago, along with dinosaurs and flying reptiles. By the time North America's earliest humans arrived in California, the once broadly distributed trees had retracted significantly, and by 10,000 years ago, after the last great ice age, there remained in America only two long assemblages of sequoia descendants, both in what would eventually be called California. One hugged the foggy Pacific coastline and the other preferred the sunny western slopes of the Sierra Nevada. These giants thrived, without humans, for a very, very long time, and so a complete account of their existence would make for a very different book. Even a comprehensive history of just human interaction with sequoias would need to cover several thousands of years, and although we have some stories passed down through the generations of native groups who lived and traveled along the lengthy band of giants, we have lost most of the experiences, tales, and wisdom of the first peoples to have walked in the shadows and sunlit reflections of these mountain forests.

What we do have—that which can be verified with historical writings, drawings, paintings, and photographs—is a story less than two centuries old. But what a story it is! In this book, Bill Tweed takes us on a thorough and richly documented tour, beginning with Euro-Americans' first encounters with the trees and the subsequently convoluted, if not laughable, competition to name this "newly discovered" species. Spanning the early days of Yosemite and Calaveras through the modern era, his story leads us not only through 250 linear miles of the Sierra but also throughout California, around the Horn, back to the East Coast, well into Europe, and even around the world to New Zealand. Bill's tour introduces us to scientists, scoundrels, and entrepreneurs, to artists and altruists, and to philanthropists and politicians. It teaches us something about history, art, science, and literature, and much about ourselves. Many a familiar name will play on the sequoia stage: Muir, Clark, Roosevelt, Olmsted, Grosvenor, Mather, Rockefeller, Clinton, and both Bush presidents.

This story also details the horrific injustices exacted upon these trees in the late nineteenth and early twentieth centuries: excavations, tunnels, slices, and slabs removed out of simple curiosity and for short-term profit. Mammoth trees, 2,000 years old, were skinned alive and left to die. Entire groves were slaughtered and carted away for grape stakes. Contrast this with the giant sequoias that now co-reign, with the coast redwood, as the California state tree, that are protected and revered in state and federal reserves throughout the Sierra, and that proudly adorn the uniforms of national park employees across the nation.

Bill expertly links the story of the trees to the evolution of the American experience with nature, including the birth of the National Park Service and of the conservation movement as a whole. This intimate story—of Big Trees and us—will persevere as a lasting acknowledgment and tribute to wild things, wilderness, and our own development as a species that can protect as well as destroy—the two sharp edges of our very powerful sword.

Bill Tweed is uniquely qualified to tell the evolving story of the Big Trees, having worked in the National Park system for thirty-two years, most of them in Sequoia and Kings Canyon National Parks, where he was charged with helping to protect many of the most precious groves of Big Trees on the planet. Until his recent move to Oregon, Tweed resided in Three Rivers, California, a Tulare County village that advertises itself as "the Gateway to Sequoia and Kings Canyon National Parks." This book was clearly written by someone from *within* the forest—someone who has spent an enormous amount of time exploring and observing, questioning and thinking about human relationships with the Big Trees. It would be difficult to find a person alive today who knows more about giant sequoias than William Tweed.

My old Park Service leather belt is dry and cracked, but it still holds a special place in my dresser. And my heart will forever hold memories of my experiences as a ranger and instructor in the Sierra. I'm reminded that our daughters used the term "field trip" long before the word "vacation" entered their lexicon, and I hope to pass that perspective down to the next generation. These days, my wife and I travel with our grandchildren, wandering from one American natural treasure to another, both teaching and learning as we go. Reinvigorated after reading this book, I am excited to share Bill's stories with the younger generation, and I am already planning trips to visit most, if not all, of the groves of Big Trees in the Sierra. There we will continue to contemplate our relationship with these and all other wild species that live protected, hopefully forever, by a park system greatly inspired by *Sequoiadendron giganteum.*

Joe Medeiros
Editor-in-Chief
Sierra College Press

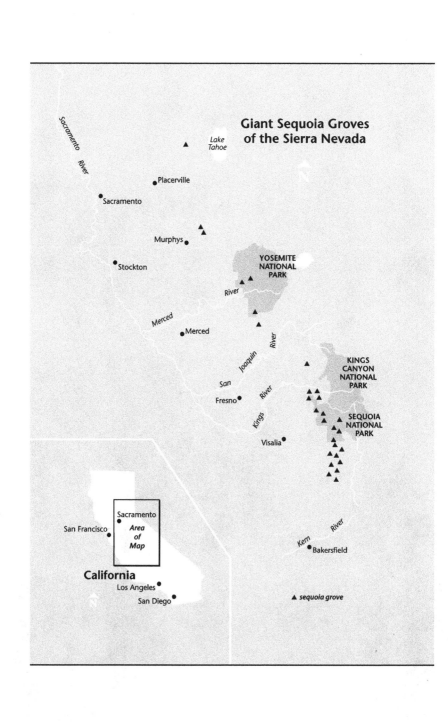

Introduction

They come for many reasons, but come they do—a million each year. Like religious pilgrims approaching a shrine, they arrive on foot, and as they arrive, most simply stop and stare in awe. Before them towers the largest tree in the world.

The giant sequoia tree known since the late nineteenth century as the "General Sherman" rises boldly into the forest canopy of California's Sequoia National Park. The weathered, maroon-colored trunk of this largest of trees leads the eye upward in the same way that a tall building pulls our attention toward the sky. Massively perpendicular, the General Sherman's bole soars upward. In human terms, one has to follow it skyward the equivalent of a dozen stories before the first significant branch interrupts its imposing bulk. Higher up, man-sized branches and foliage eventually clothe the trunk, obscuring its upper structure. An interpretive sign provides the answer that escapes our senses. The tree, it tells us, is 274 feet tall.

The gigantic base, separated from the asphalt-paved viewing trail by heavy split-rail fencing, also challenges our ability to measure by eye. We are not accustomed to estimating the dimensions of natural objects this large, and the general absence of human structures in the vicinity means we are without the usual comparative objects that allow us to judge scale. Again, to gain some sense of the tree's mass, we are reduced to consulting the interpretive wayside. We learn that at its base the tree's diameter averages some thirty-six feet, a measurement our twenty-first-century minds shallowly recognize as equal to the width of three freeway lanes. The comparison feels insensitive, but the human world is the one we live in, and the tree challenges us to connect it to daily reality in any way we can.

But I am not here today just to marvel. Over the past forty years I've spent far more time at the General Sherman Tree than I can accurately quantify. An estimate of thousands of hours feels conservative. Decades ago, as a much younger man, I stood here day after day in the gray and green uniform of a national park ranger. Wearing my flat-brimmed ranger hat, I defended and defined the tree to all who would listen.

Defending the tree proved easy enough. The problems I faced usually involved such minor matters as visitors pocketing sequoia cones or climbing over the fence and approaching the tree either to take a photo or, less benignly, to carve a name into its exposed wood. I usually just called such folks back to the fence, explained quietly why they must remain outside its protective lines, and sent the miscreants on their way.

Defining the tree proved more interesting. Questions flooded out of those who had come to see it. "How old is the General Sherman?" "What happened to its top?" "Can I grow one of these at home?" And always, seeking the famous image depicted upon a million postcards: "Where is the tree you can drive through?"

But some did inquire more perceptively—queries that allowed me to share interesting and sometimes surprising stories. They asked about the tree's black scars, and I explained the sequoias' relationship with fire. They asked about cones, and I helped them understand the reproductive strategies of the species. They asked about age, and on good days we moved on into conversations about *how* the trees lived so long, and *why*.

In the end, all these questions circled around an inescapable fact: the Big Trees fascinate us.* The attraction is not surprising. We have always

---

* In this book we will employ a number of long-used common names for these trees, including not only "giant sequoia" but also the highly descriptive name "Big Trees."

been drawn to trees. Human cultures over many millennia have noted and celebrated trees in ways that surpass the recognition given to most other plants. It's wired into us. Our appreciation of trees may date back even to Africa and our evolutionary origins. But to explain our relationship with the sequoias we must seek more, and indeed there is much more to be found.

We humans are fascinated by superlatives—phenomena that are out of the ordinary or above and beyond the norm. We seek out the largest and the smallest, the oldest and the most ephemeral, the dominant and the excruciatingly rare. Our special attraction to the sequoias begins here, with the trees' many superlatives. In my ranger days, I used to describe to visitors how the trees attracted us by satisfying some of our most human preferences: they were very large, very old, and rare. Tongue in cheek, I sometimes called this trio of attributes "the trifecta of human interest." It is hard to imagine a tree offering more to the human psyche.

When we rangers talked about the sequoias to those who came to see them, we had no choice but to start with their size. Almost no one succeeds in imagining just how large the trees are until they see them in person. "I knew they were big," visitors would say, "but I really had no idea." In ranger talks, we catered shamelessly to this perception. We told stories of how the General Sherman would compare in height to urban high-rise buildings, of how much wood was within its massive trunk, of how many three-bedroom homes could theoretically be constructed if it were milled into lumber. Visitors listened attentively, and such descriptions almost always increased their appreciation of the size of the trees.[†]

Their age demanded our notice as well, but it was only when we talked about the age of the sequoias in human terms that we could fully engage the attention of our audiences. The fact that a tree still growing today was alive during the time of Christ had enormous power over visitors, as did the detail that the entire history of our nation might be measured in only the last few inches of growth rings in the trunk.

[†] Now that I've captured your interest, the General Sherman Tree is roughly the height of a twenty-eight-story building; its trunk contains about 600,000 board feet of lumber—enough, by traditional measurement, to construct more than fifty five-room homes.

Even as we rangers sought to be science teachers, however, our visitors sought other values. Over time, we learned that most of them came not to learn about trees themselves but rather to experience something they could not otherwise imagine. The story, we were reminded again and again, was not so much about the trees as it was about the people who came to see them, and how those people reacted to what they found.

Spend enough time among the Big Trees, and core human stories come into focus. The enormous size and age of the sequoias surpass daily human experience in every way. They dwarf us, place us in perspective, remind us that we are only a tiny part of creation. They present us with that most rare of experiences: a feeling of awe. In this deeply emotional way, the Big Trees serve as a catalyst for much that is both beautiful and mysterious. By taking us beyond ourselves, they transport us into the larger world that surrounds us. They challenge us to consider the complexity of existence, the purpose of life on earth, and, for many, God. It is these complex emotional connections — even if not fully expressed by many of the people who feel them — that give the trees their enormous power not only to fascinate but also to inspire and motivate.

To understand the significance of the Big Trees, I eventually realized, one needed to explore not only botany and forest management but also history, literature, politics, philosophy, and even religion. In the following decades, as my national park career advanced well beyond tree guarding, I pursued this realization and the complexity it implied. Now, I'm returning once again to the puzzle of the sequoias and their ability to affect us. In place of the enthusiastic young ranger who once stood here, I now write as someone who has lived, worked, and pondered the sequoias for more than forty years.

As long as humans and the Big Trees have coexisted, they have influenced human thought, and to this day they continue to provoke and motivate us in countless ways. The trees have become, perhaps unexpectedly, part of our modern culture — arboreal citizens with whom we have an enduring and unique relationship. They have shaped not only our art, literature, and philosophy but also our thoughts about science, politics, and land management policy. They have, in a way that cannot be ignored, become active participants in our intellectual, social, and political worlds.

Our modern landscapes—both physical and intellectual—show the inescapable influence of the sequoias. They have inspired us to create nature preserves that today form world-famous parks, forests, and monuments. They have motivated us to create laws and policies and to rethink and restructure them over time. The trees have both fed our insatiable demand for natural resources and ultimately transcended that demand and caused us to reconsider their destruction. They have led us to restore natural fire to our forests after a century of suppression. They have revised and reformed our views of biological time and given us intriguing windows into the history of our earth and its ecosystems. They have provided a muse for writers, photographers, and painters and been the subject of countless works of human creativity. They have attracted innumerable visitors from all over the world, and then made us reconsider the implications of what we built among the trees to facilitate that tourism. The trees have even inspired us to gather up their genes and scatter them across the temperate realms of the globe, creating sequoia stands in dozens of countries far removed from their native home in the Sierra Nevada of California.

In short, the giant sequoias have captured us. No other native tree of the American West has cut so wide a swath through the human consciousness.

Many books have by now been written about the giant sequoias, but little consideration has been given to our *relationship* with the trees and, if you will allow the thought, their relationship with us. This book aims to address that gap. My goal in the pages that follow is to explore the story of how contemporary humans and these most impressive of trees have become so profoundly intertwined in so many distinctive and significant ways.

Let us begin our journey.

William C. Tweed
Three Rivers, California
October 2015

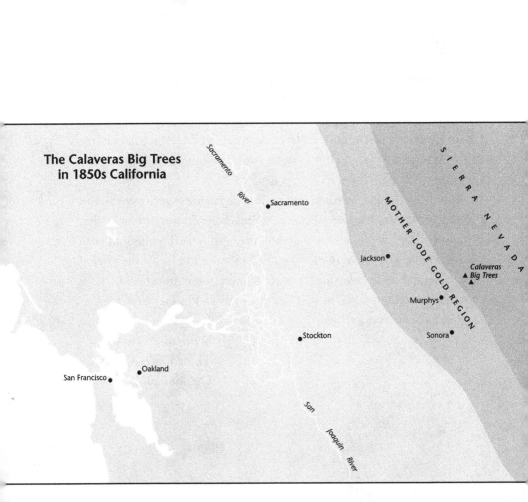

CHAPTER ONE

The Mammoth Trees of California

No longer in Sequoia National Park, I stroll down an easy earthen trail from a rustic visitor center. At first, the scene around me resembles countless other places in the Sierra Nevada—a nondescript stand of fir and pine trees. Off to the right I can see glimpses of a campground. But then two large and incongruous objects appear. To the uninitiated, they might seem a mystery. Before me a huge, platform-like stump rises some eight feet from the forest soil. Nearby, what appears to be a massive lump of wood reaches upward almost twenty-five feet, its surface deeply scored with darkened grooves. I search for the exhibit panel. There must be a story here, and of course there is. If I am to understand giant sequoia history, this is ground zero—the forest glade where our story begins.

It did not begin well. The name tells it all; the tree remnant before me bears the label "Big Stump." A wooden rail fence surrounds the massive object, but an opening provides access to a staircase that ascends onto the stump's smoothed crown. I climb up and inspect the blurred traces of more than twelve hundred annual growth rings beneath my feet. Now that I'm on top of the immense wooden corpse, I can see that the adjoining mass is the butt end of what was once an enormous log whose diameter matches the natural platform beneath my feet. I try to imagine the tree that once stood here. These are the mortal remains of the "Discovery Tree," the site where Euro-Americans took their first enduring look at the world's largest trees.

Today we call this place Calaveras Big Trees State Park. Not established until nearly eighty years after Euro-American explorers came upon the trees, the park preserves not only the sequoias within it but also the location where the opening chapters of the giant sequoia story took place. I am very near the northern end of the giant sequoia range; of the sequoia's sixty-some natural groves, only one small grove grows to the north of Calaveras.* It is no accident that our tale begins here.

The "discovery" of the giant sequoias by Euro-Americans resulted directly from the huge influx of people into California's Sierra Nevada during the gold rush of the late 1840s and early 1850s. In the Spanish and Mexican periods that preceded the boom, the few European settlers in California had left the Sierra Nevada mostly to its native inhabitants. These people—known to us today as the Miwok and the Western Mono—knew the sequoias well and spent warm summers in the Big Trees just as we do today. They perceived the trees as they did everything else that surrounded them: as part of a complex web of existence that had spiritual value embedded in every element. We should begin our story by exploring in depth their thoughts about the sequoias, but we can not, for the tragic reason that the Native American holocaust of the gold rush pushed these people to the margins and left much of their history and culture discarded and ignored by the newcomers who descended on the Sierra Nevada in the middle years of the nineteenth century. In short, as with nearly everything else they found in California, Euro-Americans listened little to those who were here before them and essentially started from scratch in building their understanding of the sequoias. So, at least for the purposes of this

---

* Since it first became apparent that the giant sequoias grow only in limited areas within the mixed-conifer forest of the Sierra Nevada, it has been traditional to call these places "groves." The exact number of groves depends upon how one defines them and whether one chooses to "lump" or "split" when it comes to sequoia stands that are separated by only small zones of non-sequoia forest. In the 1930s, some lists of groves went as high as seventy-five. Dwight Willard's *Guide to the Sequoia Groves of California* (2000), arguably the best modern list, identifies a total of sixty-seven.

book, our story begins with the entry of Euro-Americans into the Sierra Nevada.

A few crossings of the range were made prior to the discovery of gold in January 1848, and as early as 1833 one of these parties had accidentally walked into a sequoia grove. That year, a group of fur trappers led by Joseph Walker were feeling their way blindly across the mountains when they found themselves in the presence of "some trees of the redwood species, incredibly large—some of which would measure from 16 to 18 fathoms round the trunk at the height of a man's head from the ground."[1] Traditional interpretions of Walker's route envision that the group crossed the Sierra immediately north of Yosemite Valley and thus encountered the sequoias at one of the two small Yosemite National Park sequoia groves near Crane Flat, but a much better case can be made that Walker, together with his several dozen companions and two hundred horses, worked his way through the complicated country at the headwaters of the Stanislaus River and then descended westward along the gentle uplands that parallel the north side of the Stanislaus River canyon, a route that would have led the party directly into what is now called the North Grove of the Calaveras Big Trees. Whatever the route—and it is still debated—it made no difference at the time. Although Zenas Leonard's account, including his mention of the Big Trees as quoted above, was in print by 1839, it had no enduring impact. Quite simply, no one noticed.

Hence, in the practical sense, the sequoias remained unknown to the larger world until the early years of the gold rush. The summer of 1848 brought the first prospectors into the Stanislaus River watershed, and one of them, John Murphy, established a camp that soon turned into the placer mining town of Murphys. Because the camp was located high above the canyon of the Stanislaus River, the miners at Murphys had to work hard to divert sufficient water to carry out their water-intensive work. The solution was to capture stream flow higher in the mountains and bring the water to Murphys via a ridgetop ditch. This project led to the discovery of the Big Trees.

The story of this first encounter with the Calaveras Big Trees grove has been told many times and remains a classic gold rush tale. Augustus T. Dowd, a hunter providing meat for crews working on a ditch during the winter of 1852/53, wandered accidentally into this North Grove of the Calaveras Big Trees. As luck would have it, the first tree he encountered, standing on the western edge of the sixty-acre grove, was the largest in the area: the monarch sequoia that we remember as the Discovery Tree. Dowd had a reputation as a storyteller, and it took some effort on his part to convince his friends that his descriptions of giant trees were more than just his latest tall tale. He persevered, however, and soon Dowd's big trees became a feature of local interest known as the Calaveras Big Trees—the grove taking the name of the newly formed county in which the forest stood.[2]

Word spread quickly of the freakish giant trees east of Murphys. In San Francisco, the *Daily Placer Times and Transcript* published a story on May 28, 1853, that described "a cedar tree in Calaveras County 30 feet in diameter at the butt and 20 feet in diameter at a heighth [sic] of 100 feet from the ground."[3] By the end of July, the news was being reprinted in England.[4] As these stories spread, the world began to hear for the first time of these unprecedentedly large trees in California.[5]

Here the classic gold rush story usually ends, but what occurred next sets the stage for much that would happen in coming decades to the sequoias of the Sierra Nevada. In the spring of 1853, the gold rush culture of Murphys, California, reflected the town's recent origins. The rush had brought more than 100,000 men to the California goldfields, and they had come to get rich. God had put gold in the Sierra for their benefit, they believed, and it only made sense that whatever else they found would also be theirs for the taking. Several of the men involved in building the Union Water Company ditch the previous year had noticed the ability of the Big Tree story to generate human interest, and with money on their minds, they wondered how this curiosity might be turned into a source of wealth.

They came up with several ideas. On July 19, 1853, William and Joseph Lapham filed a preemption claim on 160 acres within the grove, hoping the action would eventually give them title to the recently discovered forest. In effect, they filed a mining claim on the Big Trees. Extending this commercial beachhead, William Lapham opened a simple hotel at the site, and brother Joseph sketched what became the first commercial image to show the sequoias, which the Laphams called "Arbor Vitæ."[6] The resulting lithograph showed a Big Tree standing alone in open ground with tiny human figures at its base. A Native American encampment and a top-hatted tourist party filled in the foreground. Small print at the bottom of the poster described the tree: "Diametre [sic] 34 feet at the base—circumference 96 feet—height 290 feet—3000 years old."[7] A close-up sketch of a cone and a sprig of the tree's foliage completed the image. Britton and Rey, San Francisco Lithographers, published the image.

Meanwhile, an effort was made to cut a section from a Big Tree, the goal being to display it in cities of lowland California and the eastern United States…for a fee. The impetus for this idea came initially from a Mr. Lewis, one of the first to accompany Dowd to the Discovery Tree the previous summer. Knowing something of the size and weight of the trees, Lewis knew that he could not remove a full section of tree, so he came up with the idea of stripping a tree of its bark and turning the reassembled pieces into an exhibit. Lapham's preemption claim foiled Lewis's scheme, but the idea was then taken up by William Lapham's friend, Captain William H. Hanford, who bought rights to the Discovery Tree from Lapham and set out to bring it down.[8] To Hanford, as to the Laphams, the tree stood before them as just another gold rush asset to be mined.

The destruction of the Discovery Tree, so that its wonders could be displayed elsewhere, established a model that would haunt the sequoias for the remainder of the nineteenth century. It also proved difficult to accomplish. No one had ever felled a tree of such size before. Hanford and his friends really had no idea how to do it, but with the help of the miners that lived in and about Murphys, they worked out a scheme. First, they stripped the tree of its bark to a height of about fifty feet, a task that proved easy enough.[9] Then, determined to bring the entire tree to the ground so that they could extract a cross section, they attacked it with pump augers—long metal drills roughly two inches in diameter that were used to convert pine logs into wooden water pipes. Using these ungainly tools, a crew of five began drilling holes into the base of the towering trunk with the aim of weakening it sufficiently to cause it to fall. They drilled dozens of holes, but the tree still stood. They then moved in with saws and wedges. Finally, after twenty days of drilling and three days of wedging, the tree came crashing down on June 27, 1853, the first of many sequoias to fall to modern man, and the only one ever felled in this particular way.[10]

No one had ever felled a tree of this magnitude before, and those who witnessed its fall came away deeply impressed. The *San Joaquin Republican* captured the moment:

> O, what a fall there was! Its descent is represented as having been terrific to the spectators, and, indeed, it requires no exertion of the imagination to conceive the effect of that tall mass, which seemed to thrust its branches into the clouds, when, tottering a moment from its firm perpendicular, for the first time in 3,000 years, it wavered and came rushing through the air...till all lay prone to the earth together in one mass of helpless ruin.[11]

No one recorded any doubts about what had just been done.

Once the tree was on the ground, the crew sawed off a two-foot-thick cross section from the massive log and loaded it, as well as the ring of previously removed bark, onto wagons for transport to the San Joaquin Valley town of Stockton. From that river city, the material that became the first giant sequoia exhibit traveled on by steamboat to San Francisco, where Hanford planned to set it up and charge admission to see it. As erected within a hall near the corner of Bush and Montgomery Streets, the "Mammoth Tree" exhibit took the form of a bark-confined room some twenty-five feet in diameter and twenty feet high with a sequoia-wood floor. Inside the room, Hanford set up thirty chairs.[12]

Once all this was assembled, Captain Hanford began marketing his creation. The first showing, a private event, occurred on September 2, 1853. This attracted the attention of the newspapers, which was exactly his intention. Among others in attendance were apparently several members of the California Academy of Natural Sciences and British plant collector William Lobb, who will later play a larger role in our story.[13]

"The Giant Tree," as it was called that autumn, remained on display in San Francisco for about two months. For the first several weeks, admission cost a dollar. Once the full-fare audience had been exhausted, prices were reduced, and the curious could enter and enjoy the novel scene for a tariff of fifty cents. Initiating a theme that endures to this day, the newspaper advertisements for the exhibit emphasized statistics that documented the immense size of the tree:

Look at its dimensions and judge for yourself:

Entire height unbroken	350 feet
Girth at the base	96 feet
Dia. 200 feet above the base	46 feet
Dia. 250 feet above the base	12 feet[14]

Continuing in this same vein, the notice went on to certify that the original log had contained 600,000 board feet of lumber and that it weighed 1,700 tons. And, if all that was not enough, the room within the tree was also available each evening for cotillion dancing.[15] Modern interpretive exhibits still display similar language.

By November 2, workmen had begun dismantling the exhibit and packing it into forty-three large wooden boxes for shipment around Cape Horn to New York. The cargo left San Francisco on December 31 on board the 230-foot clipper ship *Hurricane*. Captain Hanford sailed on the same day but on a steamship headed to Nicaragua, a routing that would bring him, via connecting Caribbean service, to New York well in advance of his Big Tree cargo, which sailed all the way around the southern tip of South America. Hanford arrived in New York in the second week of February, almost two months ahead of the anticipated arrival of the *Hurricane*. This gave him time to find an appropriate venue for his Big Tree show.[16]

Back in California, Hanford had made vague statements about showing his Big Tree exhibit at the world's fair that had opened the previous summer at a venue known as the Crystal Palace. Once in New York, however, Hanford realized that this would not work. Plans for the fair's 1854 season emphasized practical examples of America's "industry and skill"— for example, displays that featured new mechanical inventions and American art—and shied away from "monstrosities" like oversized tree parts. So Hanford instead arranged an independent exhibition, which opened on May 30 at 596 Broadway.

In the meantime, however, while Hanford was awaiting the arrival of the *Hurricane* and renting his exhibit hall, affairs in New York took an unexpected turn. Much to Hanford's surprise, a rival display appeared. This tree, also from California, was not a genuine Big Tree from the Sierra Nevada, but it nevertheless presented itself as just that. The display, taken from a California coastal redwood growing in the canyons on the eastern side of San Francisco Bay, took the form of a hollowed-out, but still-intact, section of tree about twelve and a half feet in diameter and eight feet long.† The log had sailed east lashed to the deck of the clipper ship *Messenger*, which had sailed into Philadelphia in late January 1854. Capitalizing on the publicity already generated about the mammoth trees of the Sierra Nevada, the hollow log's owners found it expedient to present it as a sample of that new species. This rival tree opened in a tent in Philadelphia at the corner of Broad and Locust Streets, and then, in early May, it was relocated to 472 Broadway in New York City, a location barely a block away from where Hanford was setting up.[17]

Both displays opened in May, with the Calaveras sequoia being the first and the rival tree coming in about ten days later. Both made their cases for attention through advertisements in newspapers and broadsides posted around the city. Hanford touted that his was the authentic original California Big Tree, and that there were no others. The competing tree emphasized that it was intact, not "pieced up or manufactured." The two commercial endeavors fought it out over the summer of 1854, and although neither exhibit advertised a closing date, they were both closed by the end of August.

Hanford had mentioned plans for taking his tree exhibit on to Europe, but at the end of November he sailed home to California, leaving his exhibit materials behind. The situation became clear the following spring, when an entrepreneur named John Tryon surfaced as the display's new owner. Tryon made arrangements with the City of New York in March to display the hollow-tree room outdoors in a city park, and he talked of taking it to Paris for the summer. This did not happen, however, and it appears

---

† The California or coast redwood (*Sequoia sempervirens*) is the giant sequoia's closest relative. The trees grow naturally in the mountains along the coast of California from south of Monterey to just north of the Oregon border. In this range, they are common. Although large specimens of this species are not as large as the giant sequoias in total mass, the coastal redwoods hold the title of the tallest trees in the world.

that instead the tree parts went into storage, where they were destroyed by a fire the following December.[18] We can only wonder whether anyone noticed the irony that a tree that had endured for more than 2,000 years in the wild ended up as a commercial exhibit that lasted barely two years.

As short-lived as it was, however, the Big Tree exhibit war fought on the streets of New York City in the summer of 1854 brought enormous publicity to the mammoth trees of the Sierra Nevada. Both exhibitors utilized the city's media to attract visitors, and both produced booklets for sale. These two publications, the first ever issued about the giant sequoia, focused on the story of the Discovery Tree and its preparation as an exhibit. There was more than a little irony in this as well, since the first of the two booklets to come out was sold at a site featuring what was really a coastal redwood log. The booklet issued by the misrepresented redwood log is nevertheless recognized as the earliest published work about the giant sequoia outside of newspapers.[19]

Meanwhile back in California, not everyone approved of the destruction of the Discovery Tree, and despite his ambitions, the undertaking had brought no significant wealth to Captain Hanford. Yet even while Hanford displayed his tree on Broadway, another crew went to work in the Calaveras Grove with similar intentions. Entrepreneurs George Trask and George Gale had learned something from the felling of the Discovery Tree. From Lapham, the pair purchased rights to another Calaveras Grove tree, called "Mother of the Forest," for $1,000. Instead of felling the monarch outright, however, their method involved encasing the tree in scaffolding up to a height of 116 feet and then proceeding to remove the tree's bark from the standing trunk, an effort that took a full ninety days. This fatal flaying of the tree completed, Trask and Gale crated their sixty-some tons of collected materials and shipped them to San Francisco, where they also engaged a ship headed for New York.[20]

Filled with the stripped bark of the Mother of the Forest, the clipper ship *Thomas Wattson* sailed from San Francisco on January 13, 1855. After an average-length voyage of 127 days around Cape Horn, the *Thomas Wattson* arrived in New York harbor on May 24. Within a few days Gale and Trask removed their cargo from the ship and began assembling it for

During the summer of 1854, exhibitors erected scaffolding around the Mother of the Forest Tree at Calaveras Big Trees and removed its bark. The tree died soon thereafter, but the deteriorating scaffolding remained in place for decades.
COURTESY OF CALAVERAS BIG TREES ASSOCIATION

THE MOTHER OF THE FOREST (*Sequoia gigantea*), CALIFORNIA.

display at New York's Crystal Palace, the very place Captain Hanford had hoped to use the previous summer. The hall had changed hands and was now available for more-commercial exhibits. The light-filled central hall of the Crystal Palace rose to 118 feet, and it seems highly likely that Gale and Trask had had this venue in mind from the beginning, since the bark exoskeleton they had removed from the Mother of the Forest rose to 116 feet.[21]

Initially, Gale and Trask set up only the bottom 28 feet of their hollow bark cylinder, and on Independence Day more than 6,000 patrons paid to see the exhibit. A few days after the holiday, the remainder of the exhibit was put together, thus raising the tree trunk all the way to the glass ceiling. The great column of the Mother of the Forest remained there throughout the summer.

There was considerable discussion in the newspapers about whether the bark on display actually came from a single tree or was instead an elaborate hoax. On August 9, the *New-York Daily Times* confessed that it thought the whole affair might be a sham—a "manufactured" marvel rather than a product of nature.[22] In response, Gale and Trask collected and published affidavits from several well-known persons, including Alvin Adams of the Adams Express Company and California senator William Gwin, both swearing that they had seen the tree standing at Calaveras and that it was genuine.[23]

Gale and Trask closed their exhibit at the Crystal Palace on November 1 and packed it up for shipment to Europe. At this point, Gale withdrew from the enterprise, and George Trask moved forward as the solitary operator. By mid-April 1856, Trask had transported his Big Tree exhibit across the Atlantic and set it up in London. After a private viewing on Newman Street, the material was relocated to the Adelaide Gallery on the Strand for its premier public showing in Britain. After a season there, George Trask sold the remains of the Mother of the Forest for £350 to what would be the tree's ultimate destination: the Crystal Palace exhibition hall in the London suburb of Sydenham.[24] There, erected to its full 116-foot height, it remained as the North Gallery's centerpiece until both building and tree were destroyed by fire in late December 1866. Once taken into a city, it seems, these Big Trees lost much of their natural fire resistance.

After being taken across the Atlantic in 1856, the bark shell of the Mother of the Forest tree was ultimately erected in the Crystal Palace exhibit hall in the London suburb of Sydenham, where it impressed visitors for the next decade. COURTESY OF CALAVERAS BIG TREES ASSOCIATION

As for the Mother of the Forest itself, the removal of the tree's bark doomed it to inevitable death. Given the tree's robust vitality, the process apparently took several years, but by 1861, accounts confirm that the tree had become a huge standing column of dead wood. The towering trunk clearly showed the scars from where Trask and Gale had attached their scaffolding to it.

The fate of the Discovery Tree and its near neighbor, the Mother of the Forest, provide the defining first acts in the saga of the Big Trees. Many of the themes that would be played out in more depth in coming decades make their first appearances between 1853 and 1856. Here, the concept of monetizing the public's fascination with the trees—especially their dimensions and age—came sharply into focus. The arrival in New York of two giant sequoia exhibits within a year (as well as a third exhibit purporting to be from a Sierra Nevada mammoth tree) clearly demonstrated their commercial value. All three of these projects saw their giant trees much like gold mines: something valuable provided by nature to be harvested and turned to profit by the first to control it. The fact that none of the exhibit initiators made the profits they had hoped for established yet another pattern that would continue to define the relationship between humans and giant sequoias in coming decades. Here also we see the earliest round of opposition to these types of endeavors. Many who saw the exhibits doubted their veracity, and at least a few expressed outrage that the trees had been in effect murdered by their exhibitors.

The total number of people who saw these first giant sequoia exhibits was relatively small, a factor of both their display in only a handful of cities and the fact that many would-be patrons found the entrance fees beyond their means.[25] But by 1860 knowledge of the trees and their significance as the "giants of the vegetable kingdom" had spread far and wide. This fame came from a variety of sources, but the Big Tree exhibitors played a critical role through their advertising and promotional booklets. In cities where the exhibits were marketed, street posters and handouts were spread broadly through the populace, and much of this material was picked up and reprinted by publications in other cities. As a result, even if many doubted the veracity of the exhibits, by 1860 most residents of America's growing cities—and even those rural residents who read big-city newspapers and magazines—had not only heard about the mammoth trees but had also seen them represented in engraved images featured on everything from newspaper cuts to poster-sized lithographs.

All of this was true not only of the United States but also of England. The deep anonymity that had sheltered the giant sequoias since time immemorial had been burned away by elements that defined the time: fame and greed.

After selling the Mother of the Forest to Trask and Gale, William Lapham resolved to focus on tourism rather than further tree destruction. He had opened simple accommodations on his site in 1853, and by the summer of 1855, Lapham and his partner, A. Smith Haynes, were operating the Big Tree Cottage close by the ruins of the Discovery Tree. A lithograph from that summer shows a two-story, L-shaped wooden structure.[26] Immediately adjacent to the cottage, the stump of the Discovery Tree had been covered over with a pavilion made of tree boughs.

The following year, Lapham and Haynes erected a much grander hotel, a structure that would endure until 1943, when it was destroyed by fire. They placed this new hostelry, the Mammoth Grove Hotel, some two hundred yards from the site of their first simple accommodations. This put the two-story white-painted structure outside the grove but within clear view of the Big Trees. As completed, the capacious front porch of the hotel looked down across a fenced lawn and a marble fountain to the felled Discovery Tree and its surviving siblings. The crude original hotel was removed, and a twelve-sided, wood-framed, dome-roofed pavilion was erected atop the Big Stump. Inside the pavilion, the smoothly planed stump, with its 1,244 annual growth rings now visible, became a floor suitable for dancing. Next door, a bowling alley was laid atop the fallen length of the Discovery Tree log. With the grand opening of the hotel in August 1856, the era of formal Big Trees tourism began.

Lapham and Haynes seem to have overextended themselves in the construction of their hotel, however, and they lost control of the property in a tax lien sale in 1857. By 1859, the property had fallen under the control of James Sperry and John Perry, hoteliers from the nearby town of Murphys, and it was these two who moved forward to market the site and make it famous. Reflecting the change in ownership, Lapham's Mammoth Grove Hotel now became known as Sperry's Hotel.

By the end of the 1850s, the advertising of California as a premier tourist destination had begun in earnest. A decade had passed since the rush of 1849, and California had since developed real cities and institutions, and had entered the union as the thirty-first state. The region had been largely explored and its features identified, and the wonders found there provided rich grist for James Mason Hutchings, California's first tourism publicist, who began publication of his *Hutchings' Illustrated California Magazine* in July 1856. It was only a matter of time before Hutchings got around to printing a story about the Calaveras Big Trees, and his edition for March 1859 did just that, featuring an article by Nathaniel P. Willis titled "The Mammoth Trees of California."[27]

Willis begins his several-thousand-word-long essay with a detailed account of Dowd's discovery of the Calaveras Grove. It is thus to Willis that we owe the details of this story, which was then only seven years old. Later scholars have come to doubt some of what Willis recorded, but much of what he wrote has endured to this day, including his report of how Dowd had to tell his friends in Murphys to come see a very large bear he had shot, because that was the only way he could get them to follow him to the grove of unbelievably large trees.[28]

Willis's narrative continues, taking the reader on a visit to the grove that must have reflected the experiences of many others from this era. The journey begins in Murphys, where the prospective Big Tree tourist mounts a good mountain horse and begins his ride to the grove. After a fifteen-mile excursion through the conifer woods, our visitor arrives. The "gracefully curling smoke from the chimneys of the Big Tree Cottage" welcomes him to the edge of the sequoia forest. Here, upon first sight of the Big Trees, "one thought, one feeling, one emotion, that of vastness, sublimity, profoundness, pervades the soul."[29]

After a meal at the cottage—"inasmuch as it is not always wisest, or best, to explore the wonderful, or look upon the beautiful, with an empty stomach"—the tour continues.[30] Willis provides a general description of the grove and its "103 trees of goodly size." After lunch, our visitor inspects the nearby

Big Stump, the surface of which is "perfectly smooth, sound, and level." He learns that during the recent Independence Day celebrations, thirty-two persons danced four sets of cotillions upon its surface "without suffering any inconvenience whatever." Having introduced the stump, Willis now shares the story of the felling of the Discovery Tree in 1853, which he describes as a "desecration." He remarks, also without much enthusiasm, that the upper part of the prostrate tree has become a double bowling alley. The literary tour continues with a visit to the Mother of the Forest, where workmen have erected a spiral staircase up the now-dead trunk, into which, Willis notes, many visitors have already carved their names.

Willis carefully outlines the dying tree's dimensions, then wanders on, visiting the many named trees: the Husband and Wife, the Burnt Tree, Hercules, the Hermit, the Old Maid, the Old Bachelor, the Guardian, the Mother and Son, Uncle Tom's Cabin, the Pride of the Forest, the Two Guardsmen, and the Three Sisters. Finally, he wraps up his essay with some speculations about the age of the trees — up to 3,000 years old, he guesses — and reports that other Big Trees groves are now being discovered. Obviously drawn from life, the nine illustrations in the account show not only the trees as they stood then but also the felling of the Discovery Tree, the bowling alley constructed upon its fallen log, the tourist staircase attached to the vertical skeleton of the Mother of the Forest, and the Big Trees Cottage built to house curious visitors.

A modern reader of Willis's 1859 essay cannot help but be struck by how much that early effort both anticipated and set the stage for the flood of tourist literature that would follow over the next century. The discovery story, the trees' dimensions (including an approximation of how many board feet of lumber they contained), our interest in giving them individual names, the ability of the trees to create a sense of the sublime — all these interpretations persist into our own time as aspects of popular culture associated with the sequoias. Even the urge to carve our names into sequoia logs endures.

By the early 1860s, with the help of Hutchings and others, the Calaveras Big Trees had become one of the must-see destinations of California. The display in the eastern United States of the materials taken from the Discovery Tree and the Mother of the Forest, and the attention paid the trees by scientists and various popular publications, all contributed to the spreading message that California's Sierra Nevada contained the biggest trees in the world. Supplementing these other media were visual images that captured the grandeur of the sequoias.

Photographers first captured images of the Calaveras Grove in the mid-1850s, but the printing technology of that day required that these, as well as those generated by artists, usually reached the public in the form of engravings printed in magazines and on lithographed posters. Of particular significance in publicizing the Calaveras Big Trees were the lithographs issued in San Francisco by Edward Vischer. A California artist and merchant, Vischer published his *Vischer's Views of California: The Mammoth Tree Grove and Its Avenues* in 1862.[31] *Vischer's Views* contained twenty-five engravings that focused, despite the expansive title, on Calaveras County, California. Containing not only engravings but also text, the publication took the form of a guidebook, the first to feature the Big Trees.

The oval-framed frontispiece compressed much of the grove's history into a single image: the Big Stump and the prostrate remains of the Discovery Tree filled the center of the scene, while the scaffold-framed and obviously dead Mother of the Forest towered in the background. After introducing six scenes from the mines and towns neighboring the grove, most of the remainder of the volume centered on the Big Trees. In addition to several views that showed the hotel and its grounds, other plates featured the trees themselves, including images of the Father of the Forest and Hercules trees. Particularly powerful was Plate IV, which displayed a double exposure of the Mother of the Forest—an 1855 version showing it still alive, and an adjacent 1861 version presenting the tree as dead and leafless, its trunk marred by the tourist staircase. To emphasize the fragility of the trees, Vischer placed the stump and butt log of the Discovery Tree in the foreground.[32] In all this, Vischer captured the scene in such a way that his subjects are still recognizable even a century and a half later.

Interpreting the meaning of Vischer's Big Tree images from a twenty-first-century perspective can be a tricky business, however. Some cultural historians have proposed that Vischer's artistry is best considered as an artistic commentary upon the fragile condition of the United States at

The Mammoth Grove Hotel at Calaveras Big Trees, as documented in this 1862 engraving by Edward Vischer, would become the first of many Big Tree resorts to be built in the Sierra Nevada. COURTESY OF CALAVERAS BIG TREES ASSOCIATION

the beginning of the Civil War. In this interpretation, the sequoias in their natural state represent the Union, and trees despoiled or felled for commercial display stand for the threats to the Union presented by the Confederacy.[33] Such elaborate analyses may not be necessary, however, to make sense of these images. Vischer may have been nothing more than a businessman both trying to make some money with images of Calaveras County and expressing his feelings about the destruction of two of the grandest trees in the grove. As we have seen in Willis's 1859 article for *Hutchings' Magazine*, such views were not unheard of during this era.

Whatever baggage we wish to load upon Vischer, the fact remains that his work captured the attention of the masses and helped publicize the Calaveras Big Trees at the moment when they were taking off as a major attraction. The next twenty years would see the golden age of tourism for the site. Numerous magazine articles extolled the magnificence of the trees. Sperry's Hotel grew and prospered. Photographic images of the trees proliferated and largely erased those early rumors that claimed the enormous trees of California were little more than a hoax. By 1871, two years after the completion of the first transcontinental railroad, a branch railway had been pushed to within forty miles of the grove, and improved roads made the stage connection from its terminus a relatively easy journey, at least by nineteenth-century standards.

At Calaveras, by the early 1860s we find the beginnings of something easily recognizable in the twenty-first century: giant sequoia tourism. Improved access, a well-publicized hotel, named features—all these would become defining characteristics of Big Tree tourism. But, at least for Calaveras as a preeminent destination, the writing was already on the wall. One by one, other sequoia groves were being located and publicized, drawing attention away from the original site. The world of the Big Trees would turn out to be much more extensive than first suspected. New and even bigger giants awaited discovery.

CHAPTER TWO

To Name Is to Know

I recognize the tree immediately, even though it's far from its home. The massive symmetrical cone of foliage soars into the air. A quick glance at its profile convinces me that it must be a giant sequoia. It's a bit out of place, however. I'm walking through the urban heart of Queenstown, New Zealand.

The tree before me is well established in its adopted setting; it is at least eight feet in diameter at the base and over a hundred feet tall. The same label that gives me the tree's local common name also informs me that it was planted in 1874. It stands, along with several equally large sequoia siblings, in the very heart of this tourist city at the imperially named corner of Stanley and Ballarat. Locating it presents no challenge, for it towers above this generally low-rise city, famous for its spectacular natural setting and outdoor sports opportunities. The only other trees in the city that approach it in height are a cluster of tall spires in the Queenstown Gardens, the extensive city park that occupies a low ridge overlooking the glacially carved Lake Wakatipu. I seek out these other trees—sequoias, of course—and find them even larger than the Big Trees in the city center. A glance at the botanical label tells me that I am looking at specimens of "Wellingtonia," a tree from California that also bears the scientific name of *Sequoiadendron giganteum*.

The questions of how these trees got *here* so long ago, and why in this setting they bear a British imperial name, suggest that I have some homework to do if I am to understand the scientific and popular nomenclature that surrounds these arboreal giants.

From the very beginning, botanists took at least as large an interest in the Big Trees of California as did those who sought to display them for profit. Plant scientists had the opportunity to become involved soon after the discovery of the trees at Calaveras because in June 1853, Augustus Dowd, the man credited with discovering the grove, collected several branches bearing both foliage and cones and shipped them to Albert Kellogg in San Francisco. Dr. Kellogg, a physician with a strong interest in botany, had come to California in 1849 after, among other adventures, spending time naturalizing with John J. Audubon in the then Republic of Texas. On the evening of April 14, 1853, two months before Dowd sent him the tree samples, Kellogg met with six other men interested in California's natural history to form a new organization they called the California Academy of Natural Sciences. Modeled on a similar scientific organization created in 1821 in Sydney, Australia, the California Academy would become the first scientific academy in the Far West of North America.[1]

So it was to Kellogg that Dowd forwarded his samples from the trees at Calaveras. Kellogg now had in his possession everything he needed to provide the species with an official scientific name, although he chose not to do so immediately, preferring first to consult with experts and books not available in California. Kellogg also sent samples to friends and colleagues who shared his interest in the subject, among whom were John Torrey at Columbia College in New York and Asa Gray at Harvard, probably the two most esteemed American botanists of their generation.[2] In taking his time before formally christening the new species, however, Kellogg unintentionally set in motion a sequence of events that would cause arguments for decades to come.

Closer to home, Kellogg apparently also showed his specimens to William Lobb, an Englishman who in the summer of 1853 was collecting plants in California for the nursery firm of Veitch & Sons, located near the city of Exeter, in Devonshire, England. A fascinating early Victorian, Lobb had been traveling in search of plants almost continuously since 1840, when he first left Britain for Brazil. Since 1849 he had been diligently at work along the Pacific Coast of the United States, seeking interesting native plants in California and the Oregon Territory. As luck would have it, he was in San Francisco late in the summer of 1853, and there he attended the September 2 event put on by Captain Hanford to display his newly arrived mammoth tree materials from Calaveras.[3]

Over the previous several years, Lobb had invested considerable time collecting seeds from the conifer trees native to the Pacific Coast, and as soon as he heard about the Big Trees at Calaveras, he immediately understood the species' horticultural potential in the British Isles. Exotic conifers had become almost a mania in England, and now, Lobb realized, he might be privy to the discovery of the ultimate coniferous tree.

What Lobb did next clearly demonstrates how important he thought the new tree species might be. The English plant collector seems to have left San Francisco almost immediately for Calaveras, traveling to Murphys via overnight steamboat and a thirty-six-hour stage ride. From there, he moved on to the grove, where he collected foliage, cones, seeds, and perhaps even some live seedling trees. He returned to San Francisco just as quickly, arriving there in time for the next fortnightly "steamer day," when steamships left the city for Central America with connections on to Britain.[4] This sudden return to his home island by a man who had been traveling abroad for more than a decade strongly suggests the importance he assigned to his latest botanical discovery.

We do not know exactly by which routes and ships Lobb traveled eastward from Central America, but we do know that he was back in England no later than mid-December.[5] Once in Britain, Lobb not only provided horticultural materials to Veitch & Sons but also gave botanical samples of the new tree to botanist John Lindley, who, having none of Kellogg's reticence, immediately proceeded to describe and name the tree.[6] From the materials in hand Lindley concluded that the trees were sufficiently different from known species to deserve their own genus. Thus, when the first and defining botanical description of the mammoth trees was published by Lindley on December 24, 1853, in *Gardeners' Chronicle*, a London-based publication, he chose to name them after Arthur Wellesley, the Duke of Wellington. The famous British hero had died just fourteen months previously.

> We think [wrote Lindley] that no one will differ from us in feeling that the most appropriate name to be proposed for the most gigantic tree which has been revealed to us by modern discovery is that of the greatest of modern heroes. Wellington stands as high above his contemporaries as the California tree stands above all the surrounding foresters. Let it then bear henceforth the name of "Wellingtonia gigantea."[7]

As word of Lindley's publication notice spread via the slow communications channels of those times, the bestowing of a British general's name upon the great trees of California created a furor. Kellogg would insist for the rest of his life that he had told Lobb in San Francisco that he had already named the tree the Washington cedar. In May 1854, Asa Gray weighed in with talk to the American Academy of Arts and Sciences (published in the *American Journal of Science and Arts*) in which he noted that the tree "is a very close relative of the Redwood of California" and that although Lindley has assigned it the name *Wellingtonia*, "additional materials are needed to confirm the genus."[8] Meanwhile, other Americans also began to offer alternatives. Over the next several years, these included *Americus gigantea, Taxodium washingtonium, Washingtonia californica,* and *Taxodium giganteum*. This last name was bestowed in 1855 by Kellogg, belatedly weighing in on the subject nearly two full years after his first encounter with the species. Clearly, the multiple references to the nation's founding father were an attempt to counter the assignment of the name of the Duke of Wellington to a plant unique to the United States, but under the rules of botanical nomenclature, none of these after-the-fact renamings by Americans held ground.* It was, in fact, a Frenchman who came up with a replacement, and this time the alternative gained wide acceptance.

---

* In general, the first scientific name given to a newly discovered species takes precedence over all other subsequent proposals unless there is a *scientific* reason to rename the tree. An example of this latter condition would be the realization that an organism is so closely related to another genus that it should be merged into that existing grouping. As we shall see, this is what happened to the Big Trees.

We might assume that Joseph Decaisne had even less reason to appreciate the political implications of the name *Wellingtonia* than did his American cousins, but Decaisne moved forward on botanical rather than political lines. On June 28, 1854, the Belgian-born Frenchman presented his findings to the Société Botanique de France, claiming that Lindley had been in error and that the trees did not deserve to be placed in their own unique genus. Instead, he advocated, the huge trees of the Sierra Nevada ought to be recognized as close relatives of the redwoods growing along the coast of California, which had themselves been rechristened after their original name was deemed inadequate. In 1828 they were termed *Taxodium sempervirens*, but in 1847 the name was changed, by Austrian botanist Stephen Endlicher, to *Sequoia sempervirens*. Endlicher had based this successful argument on the premise that the coastal trees were too distinctive to be lumped within an existing genus of cypress trees. Instead they deserved their own genus—*Sequoia*.

Taking a fresh look at the Big Trees of the Sierra Nevada, Decaisne now proposed that they were so closely related to their coastal cousins that they should be recognized as a second species in the genus *Sequoia*. From this base, rejecting *Wellingtonia* but retaining *gigantea*, Decaisne proposed the name *Sequoia gigantea* for the Sierra's giant trees. Recognizing this as a reasonable solution, botanists on both sides of the Atlantic accepted the new designation, and for the next hundred years, most would use this name.[9]

Decaisne's assignment of the mammoth trees of the Sierra to the genus *Sequoia* solved one problem but created another, however. When Endlicher first applied the genus name *Sequoia* to the California coastal redwood, he omitted one critical detail. To the frustration of later botanists, he did not identify the source of the name. The best attempt at a botanical explanation of the origins of Endlicher's name *Sequoia* focuses on the fact that in the late 1840s the Austrian botanist was working on a taxonomic system to organize the world's conifer trees according to how many seeds their cones contained per cone scale. He had discovered a sequence of cone/scale relationships, but within that sequence there was a gap until, it seems, he studied the coastal redwoods of California. This tree filled in the gap in his sequence, and he thus named the tree *Sequoia*—the "sequence tree."[10]

In the meantime, arguments also continued over the tree's popular name. Here, because there was no protocol for resolving the question, variant names proliferated: among those used were Arbor Vitæ, great tree, mammoth tree, giant tree, big tree, Bigtree, California big tree, Sierra redwood, and many more. Eventually, popular usage settled mostly on mammoth tree or Big Tree, and these remained the most popular common names for the remainder of the nineteenth century. (The common name "giant sequoia" would not come into popular usage until the middle years of the twentieth century.) Meanwhile, the British continued to call the trees Wellingtonia.

The same efforts that led to the formal naming of the trees also introduced them to the horticultural world. As it turned out, even before Lobb took seeds back to Veitch & Sons, Scotsman John D. Matthew had collected seeds at Calaveras. By August 1853, Matthew was back in Scotland, where his seeds were planted and rapidly sprouted. The California species promptly proved itself suitable to the mild, moist climate of the British Isles, and "Big Tree mania" swept over Britain and across the channel into Europe. Planting a Big Tree from California became a must-do for wealthy landowners and their estate gardeners.

When Veitch & Sons placed their first mammoth-tree seedlings on sale in early 1854, they sold 3,000 in one day![11] That spring the nursery listed individual seedlings for sale for two guineas, something over $400 in modern currency. To sustain interest, Veitch & Sons also displayed the tree's cones and foliage in London, sold copies of Joseph Lapham's lithograph of the Discovery Tree in its virgin state, and purchased advertising space in newspapers and journals.[12] Writing a few years later, a British observer documented that Big Trees "were planted everywhere: on suburban lawns, on great estates, in triumphal [urban] avenues."[13] Soon, the trees had spread to all the more temperate parts of the British Empire, including, by 1874, the frontier village of Queenstown, Otago Province, New Zealand.

Big Tree mania in the British Isles confirmed what had already been discovered in the United States: that the Big Trees of Calaveras possessed a powerful pull on the human imagination. Whenever people heard of them, they wanted to read more about them, to touch pieces of them, to grow them, and even, if they could afford to do so, to travel to California to see them growing in the wild. Scientists were equally fascinated. Here, many recognized, was the natural monarch of the plant world.

The trees had become international celebrities.

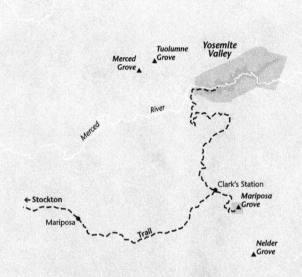

CHAPTER THREE

A Grove Called Mariposa

Another "discovery tree" rises before me, and this one is both still standing and very much alive. The majestic specimen tree rests on a gentle slope a few yards below the crest of a broad forested ridge. I study it for a few moments, then am distracted by the sounds of a slow-moving vehicle and a PA system. I turn around to face the narrow paved road upon which I have been walking, and a tractor-truck pulling an open trailer comes into view. Several dozen visitors, enjoying the open-air seating on this early fall day, crane their necks upward as they take in the scene. Speaking through a microphone, the driver/tour guide announces what I already know: that I am standing at the base of the Galen Clark Tree.

The tram moves on, and its associated intrusions fade into silence. The calls of a Steller's jay and a red-breasted nuthatch reassert the primacy of nature in this still largely wild forest. I stroll southward following the old roadway. Aside from the tram, the route seems dedicated to pedestrian use. I sense no other motor vehicles in the vicinity, however I am not walking alone. Now and then I catch sight of others—solo hikers and family groups—all taking in the great forest. In about half a mile I come to a junction. Around me towers an impressive stand of sequoias. Reading my old guidebook, I make out the

Governor's Group, the Haverford Tree, and the American Legion Tree. My 1949 guide suggests that "there are few places where one may so completely experience the sensation of being in a great outdoor cathedral, as standing here among these giant, fluted columns reaching into the vast blue vault of the sky."[1] I can only agree.

Continuing my stroll, I find myself in a lush, forested basin richly crowned with giant reddish tree trunks. In their midst, I find a long, low log cabin—the Mariposa Grove Museum of Yosemite National Park.

For a few short years, the Calaveras Grove gloried in the public eye as the only known grove of Big Trees.* Such a state, of course, could not long endure. In at least one other area, gold mining and giant sequoia groves fell within reasonable proximity, and it was inevitable that this site would become the next center of sequoia interest.

The gold mining potential of the Mariposa district came quickly into focus as the huge wave of forty-niners explored the areas north and south of the 1848 gold strike east of Sacramento. Placer miners were active along Mariposa Creek in the summer of 1849, and by November 1851 they had established a town that functioned as the official seat of the newly created Mariposa County. Over the next several years, the efforts there evolved from placer workings to hard-rock mining, and, as in Calaveras, the need to capture water to support these industrial undertakings sent engineers and others into the higher mountains to the east of the gold district.

Among those living and working in the Mariposa district was a bearded, lanky, middle-aged New Englander named Galen Clark. Born in Quebec in 1814, young Galen had moved to Dublin, Massachusetts, with his family as a boy. The years that followed remind us of the restless mobility that distinguished the North American frontiers of the nineteenth century. By 1837, Clark had moved west to Missouri, where he married, and then east

---

* The giant sequoias at Calaveras actually grow in two separate groves roughly half a dozen miles apart: the North Calaveras and South Calaveras Groves. Dowd's discoveries in 1852 publicized the North Grove, and it became the famous locale celebrated worldwide. Others encountered the South Grove as early as 1853, but despite news of its existence, the general public continued to perceive the North Grove as the sole site containing sequoias. See Lowe 2012a, 21–22.

Galen Clark first wandered into the Mariposa Grove in the summer of 1857. He would spend the rest of his long life protecting the Big Trees.
COURTESY OF YOSEMITE NPS LIBRARY

again to Philadelphia, where, in 1848, his wife died. Clark spent the next several years eking out a living in New York. Then, in October 1853, like so many others seeking yet another fresh start, Clark sailed for California. By the summer of 1854, he was searching for gold in Mariposa County.[2] To this point, there was nothing in Clark's story to suggest that he would ultimately become more closely associated with the sequoias than almost any other nineteenth-century individual, and yet that is exactly what happened.

Clark had arrived too late in the Mariposa district to make much money as an independent miner and, in the summer of 1855, he accepted employment with a company seeking to bring water to the mines from the South Fork of the Merced River. This work provided new contacts and connections, and in October, Clark joined a group of Mariposa-area miners making one of the very first tourist expeditions to Yosemite Valley.[3] To get to that not-yet-famous place, Clark's party traveled east from Mariposa, crossed over the 6,000-foot-high summit of Chowchilla Mountain, dropped into the delightful meadow country along the South Fork of the Merced River, and then crossed the ridges northward to Yosemite Valley itself. The trip changed Clark's life. Along the way he found "primeval forests" that were a "revelation" and, in Yosemite Valley, the "sanctum sanctorum of Nature's vast mountain temple."[4] The country remained embossed on Clark's mind—not only Yosemite Valley but also the meadows along the South Fork of the Merced River.

The following spring, Milton and Houston Mann, brothers living in Mariposa who has also visited Yosemite Valley the previous summer, began construction of a trail for tourists from Mariposa to the scenic canyon. The route they improved closely followed the route Clark had taken in 1855. At the same time that the Mann brothers began their project, Clark returned to the meadows of the South Fork to live, hoping to improve his health, which had deteriorated significantly over the preceding winter. On March 19, Clark filed a claim on 160 acres there, and he spent the next year nursing his lung problems with a self-prescribed regimen of outdoor living and broiled deer livers. The austere treatment worked, and his health returned. By the following year, he was living in a log cabin of his own construction measuring some twelve by sixteen feet. There, because he was living along the rough trail that led from the town of Mariposa to Yosemite Valley, Clark began to receive occasional guests. Recognizing it as opportunity knocking at his door, Clark began to assist the trickle of tourists.[5]

It was in this new role as a fledgling hotelier that Clark became interested in rumors of the mammoth trees in his vicinity. Several individuals had already stumbled across the Big Trees—one party as early as 1849—but compared to the excitement of gold mining, no one much cared. In 1855, the summer before Clark established himself at his new South Fork homestead, R. H. Ogg had again found the trees while working for the same ditch company that employed Clark that summer. Now, in June 1857, Clark and trail builder Milton Mann set out in search of the mammoth trees with the thought that such a grove might be of interest to the touring parties on their way to Yosemite Valley.

Finding the grove did not prove difficult. Clark and Mann soon located the trees on a ridge about four miles south of Clark's camp. The two had wandered about a good deal in their search, with the result that Clark first encountered the southern edge of the grove even though his cabin was to the north. He would later raise a stone cairn to mark the location from which he first saw a giant sequoia.[6] A modest man, Clark called that tree the

"Forest Queen." Not until after his death, in 1910, was the tree rechristened in his honor.[7] Today we still know it as the Galen Clark Tree.

Impressed, and more than a little awed, Clark explored the grove that summer, and it wasn't long before he realized there were actually two major areas of Big Trees—the upper section that he had first encountered, with about 260 trees, and a lower and more northerly part of the grove, with about 240 large specimens. Either unit equaled the Calaveras Grove in number of trees. Later that same year, a Native American guide led Clark south over the ridge into the Fresno River watershed, where, fourteen miles from his cabin, Clark was shown another large grove of Big Trees. Like the grove he had found earlier that same year, this new grove again had several natural segments or sub-groves. At first glance it contained about five hundred mature sequoias.[8]

Pursuing that most human of inclinations, Clark began to name the features he encountered. Eschewing the suggestion that the grove take his name, he instead christened it after the county in which it stood—hence "Mariposa Grove." He did the same with the Big Trees he discovered in the autumn of 1857; that more southerly grove became known as the Fresno Grove. Less modestly, the site of Clark's camp in the meadowlands along the South Fork acquired the name Clark's Station.[9]

We can wonder how much less attention the Mariposa Grove might have drawn had it been in a more remote location, but the reality was that the discovery of Big Trees along a trail leading to Yosemite Valley immediately connected them with that blossoming destination. Interest in the scenic glacial gorge had been growing since the first tourist visits in the summer of 1855, and with the completion of the Mann brothers' trail in 1857, the area's fate was sealed. A branch trail was soon extended into the grove from the vicinity of Clark's homestead, and Clark found himself host to a stream of visitors. By his own account: "Regular travel commenced in 1857, and I began to give entertainment at my place and as travel increased I increased my accommodations."[10] Still, the flow of tourists remained only a trickle. Perhaps seven hundred visitors made it to Yosemite Valley between 1855 and 1864, and about half of these traveled via Clark's Station. Most of these, we can assume, also made the side trip to the Mariposa Grove.

Photographs taken by Carleton Watkins in the late 1850s did much to bring the Mariposa Grove to the attention of the world. COURTESY OF YOSEMITE NPS LIBRARY

Encouraging this tourism was a marketing effort. J. M. Hutchings, of *Hutchings' Illustrated California* fame, began writing about Yosemite as soon as he initiated publication of his magazine. Hutchings filled his issues with engravings, and these images, which included views of the Mariposa Grove, began to circulate broadly. Hutchings had brought artist Thomas A. Ayres to Yosemite Valley in 1855 and commissioned him to prepare a number of drawings for the magazine's use. Over the next several years, artists James Madison Alden, James Lamson, George Tirrell, and William Smith Jewett all visited the region and created views of its wonders. In 1859, photographer Charles Weed made his way into the mountains and captured several images of the Big Trees.[11] Carleton Watkins appears also to have worked in Yosemite in 1858 or 1859 for Hutchings, and he then returned in 1861 to capture a set of Sierra photographs that ultimately helped make him famous. Among these were large-scale, glass-plate photos of the Grizzly Giant and of Galen Clark standing before a sequoia in buckskin clothing.[12]

Many of these images made their way beyond the boundaries of California. Ayres's drawings were displayed at the Union Art Gallery in New

York in 1857, and Watkins's 1861 photographs were shown in that same city in December 1862. Weed's 1859 photos, in the form of stereographs, made it all the way to England in the early 1860s.[13]

During these same critical years, Yosemite and the Mariposa Grove began to receive notable visitors whose accounts received wide publicity. Horace Greeley, publisher of the highly influential *New York Tribune*, visited the Mariposa Grove in August 1859 after spending several days in Yosemite Valley. While in the vicinity, Greeley interviewed a Mr. Wheeler about the relative merits of the Big Trees at Mariposa and Calaveras. He reported that Wheeler

> found the Calaveras trees in far better condition, in the charge of a keeper, and approached by a road over which a light carriage may readily be driven up to the very trees themselves. These are no light advantages; but he assured us that, on the other hand, the Mariposas trees are considerably more numerous (some six hundred against two hundred and fifty), and are really larger and finer specimens of their kind.[14]

Greeley went on to describe the size of the trees, the visible damage done to them by fires, and the seeming paucity of their reproduction. Impressed by their apparent fragility and rarity, he expressed the need for either the "village of Mariposas" or the state government of California to "immediately provide for the safety of the trees." Continuing on these themes, he said:

> I am sure they will be more prized and treasured a thousand years hence than now, should they, by extreme care and caution, be preserved so long, and that thousands will then visit them, over smooth and spacious roads, for every one who now toils over the rugged bridle-path by which I reached them. Meantime, it is a comfort to know that the Vandals who bored down with pump-augers, the largest of the Calaveras trees, in order to make their fortunes by exhibiting a section of its bark at the east, have been heavy losers by their villainous speculation.

All this, and much more, ran in the *Tribune* in late 1859, and then came out in book form the following year under the title *An Overland Journey from New York to San Francisco in the Summer of 1859*.

Of equal importance was the visit a year later of the renowned Unitarian minister Thomas Starr King, who, after preaching successfully in Boston for over fifteen years, accepted a position in San Francisco in 1860. That same summer, King visited both Yosemite and the Mariposa Grove, which he described in a series of letters published in the *Boston Evening Transcript*. Making the point that many were skeptical about the size of these California wonders, and admitting his own doubts, King went on to be deeply impressed by what he found.

> I lay for half an hour alone at the root of the most colossal bole—my companions out of sight and hearing—and watched the golden sunshine mounting the amber trunk, and at last leaving a hundred feet of it in shadow to flood its mighty boughs and locks with tender luster. What silence and what mystery![15]

Coming just a year after Greeley's account, the Reverend King's letters went a long way toward convincing eastern readers that the Big Trees were indeed real. Who could doubt the firsthand report of one of New England's most respected Unitarian ministers?

The cumulative effect of all this attention was that within a few short years Yosemite Valley went from being essentially unknown to one of the best-known landscapes in the American West. Such publicity, even though it was focused on Yosemite, inevitably also increased public awareness of the neighboring Mariposa Grove, a significant connection that would soon become apparent.

In February 1864, Israel Ward Raymond, California representative of the Central America Steamship Transit Company, addressed John Conness, one of California's two federal senators, with a request that Conness introduce a specific piece of legislation. Raymond's letter outlined the desired bill, which proposed that two tracts of public land in the Sierra Nevada be withdrawn from eventual sale by the General Land Office and instead be transferred to permanent ownership and management by the State of California. On the surface, there was nothing particularly unusual about the proposed transfer; in the mid-nineteenth century, states often lobbied the federal government for grants of land. What made this proposed transaction significant, however, was its purpose: the two tracts were to be set aside "for public use, resort, and recreation and [would be] inalienable forever."[16] The features that would be protected were Yosemite Valley and the Mariposa Grove.

Repeated attempts by historians to document Raymond's motives in this seminal proposal have met with failure. Whether the idea arose first in Raymond's mind or, more likely, in that of Frederick Law Olmsted, the recently appointed manager of the nearby Rancho Las Mariposas land grant, remains uncertain, but many clues point to Olmsted. Key provisions in the resulting legislation closely echo ideas Olmsted held dear, and many historians agree his fingerprints were all over this effort.[17]

By the early 1860s, Olmsted had taken major steps down the path that led him from being a journalist to becoming the father of landscape architecture in the United States. Indeed, the lessons he learned as a journalist about the importance of public spaces strongly informed his later work, which focused on public parks and university campuses, including New York's Central Park and the California campuses of Stanford and UC Berkeley. Olmsted believed that well-designed open spaces with strong natural elements were critical to the success of sustainable democratic societies—a conclusion he came to through direct experience.

In 1863, after two years in the highly stressful position of Executive Director of the Sanitary Fund (a precursor to the American Red Cross that provided medical services to those wounded on the battlefield), Olmsted stepped away from the war and accepted a position in California. He arrived in September 1863 to begin work as manager of the Rancho Las Mariposas land grant, and it is here that he becomes a part of our story.

In July 1864, Olmsted visited both Yosemite Valley and the Mariposa Grove and came away enthused by what he had found. Later that same year, when California governor Frederick Low appointed the first board of commissioners for the new state reserve, Olmsted became the nominal chairman. In this role, Olmsted drafted recommendations to the state legislature regarding the value and management of the grant that read today almost as a blueprint for the federal national park system that was developed in the late nineteenth and early twentieth centuries. Olmsted's vision emphasized preserving the lands within the reserve in as natural a state as possible, managing it in consultation with artists and scientists, and ensuring that it was accessible to all stratums of society, not just the rich. His appreciation of the Big Trees also shines through. He describes them as having "such beauty and stateliness that, to one who moves among them in the reverent mood to which they so strongly incite the mind, it will not seem strange that intelligent travelers have declared that they would rather have passed by Niagara itself than have missed visiting this grove."[18]

None of this suggests, however, that the anything-but-routine bill received more than cursory attention from Congress. Bigger things were going on in the spring of 1864. In the same month that this bill moved through Congress, federal armies fought the Battles of the Wilderness, Spotsylvania, and New Market, and General William Sherman began his March to the Sea through Atlanta. In New York, from which Raymond had written to Conness, the city was abuzz with the exploits of Pauline Cushman, an actress from Louisville who had spied for the Union and was then captured by Confederate forces and condemned to be hanged. Before the sentence could be carried out, however, troops under the command of General Rosecrans routed the rebels and rescued Cushman, who became a national heroine. Meanwhile, in the West, the Union Pacific and the Central Pacific Railroads were hard at work spiking down the tracks of the first great transcontinental line. In such turbulent times, a land grant in far-off California attracted little attention.

But the bill was significant—precedent-breaking, actually. Throughout the republic's first century, a general consensus existed that the best possible use for lands of the public domain was to sell or otherwise transfer them into private ownership so that they could be occupied and developed. Exceptions were rare, mostly for things like military reservations and lighthouses. The Yosemite Grant of 1864 did follow that model in that it transferred the lands in question out of federal ownership, but in this case it then placed them under state control with the specific instructions that they were to remain available for public use and inalienable forever. Many historians recognize this as the first clear legal expression of what would later become the national park idea.

By 1864, the American experience with the giant sequoias of the Sierra Nevada had stretched to a dozen years. Much had been learned. The giant trees' ability to inspire and capture the imagination had been clearly demonstrated not only in the United States but also abroad. At the same time, the very off-site exhibitions that had done so much to make the trees famous had also garnered considerable complaint. Many doubted that the trees were genuine, while others protested the destruction required to extract the display pieces from their natural environments. Conness brought up these points on the Senate floor as he argued for the Yosemite bill. Speaking of the tree display in the Crystal Palace, then in Sydenham, England, he reminded his fellow senators that "the English who saw it declared it to be a Yankee invention, made from beginning to end; that it was an utter untruth that such trees grow in the country; that it could not be."[19] "The purpose of this bill is to preserve one of these groves from devastation and injury," he continued, adding, "The necessity of taking early possession and care of these great wonders can easily be seen and understood." In other words, Americans had a responsibility to protect these planetary marvels; the nation's honor was on the line.

It is easy to underestimate the role the Big Trees played in the Yosemite Grant. The setting aside of Yosemite Valley for public benefit has so captured the public mind over the years that we tend to forget that the sequoias were as important a part of the grant as the famous glacial valley itself. Barely a dozen years after their "discovery" by Euro-Americans, the giant trees had significantly changed in status, no longer mere victims of human intention or mere objects of tourism. Now, their presence began to shape public affairs.

The 1864 legislative protection of the Mariposa Grove can be seen as a direct response to what had happened at Calaveras. At that first grove, the trees had rapidly gone into private ownership and been subjected to a long string of indignities and insults. Two large specimens had been killed outright. The Yosemite Grant offered a new dream. At the Mariposa Grove, the four square miles that contained the trees were to be held "inalienable" by and for the benefit of the people. These trees would not become exhibit logs in faraway cities.

The sequoias of the Sierra Nevada had worked their magic upon the body politic of the United States for the first time. There would be more to come.

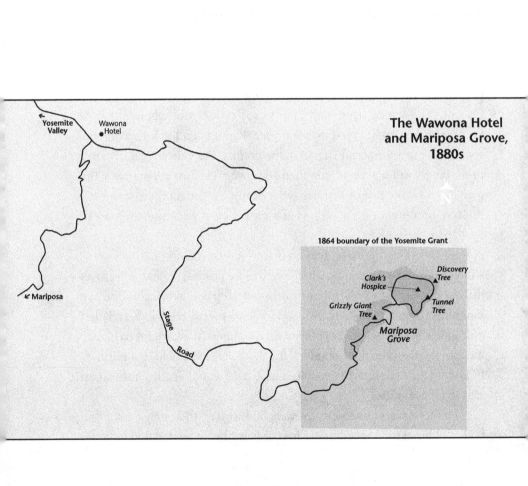

CHAPTER FOUR
An Arboreal Mecca

Somehow, these giant sequoias look at home even though nature did not put them here. They tower over the veranda-framed Victorian hotel with a confidence that makes it clear that they are defining elements of this place. Actually, however, very little about the scene is natural, especially considering that I am in a national park. A manicured green lawn rolls up the gentle slope to the front steps of the boxy two-story hotel, the turf's smooth progress broken only by an old-fashioned cobble-framed fountain and reflecting pool. The hotel itself clearly shows its nineteenth-century origins. It gleams cleanly in the bright mountain sunshine, dark green trim against white to emphasize its symmetrical exterior. The row of four Big Trees rises behind the building, their narrow, pointed spires dwarfing the building as they reach up into the blue Sierra sky.

The Wawona Hotel of Yosemite National Park endures in the twenty-first century as a monument to the power the Big Trees had to attract tourists during the last third of the nineteenth century. Nowhere else in the Big Tree country of the Sierra Nevada has a hotel of this age endured into the current era, and the continued existence of this one represents a small victory over time. Most of the wooden resort hotels of the nineteenth century eventually fell prey to fires, and of those that escaped burning, the specter of modernization took down most of the rest. But the Wawona Hotel still stands.

I climb up the wooden steps onto the veranda that circles the building. Old-fashioned benches and rattan chairs look out over the lawn and inspire leisure. I step inside and find myself in a Victorian lobby. Several sitting rooms decorated with period furnishings, paintings, and photographs invite me to pause, but I press on through the back doors and look up the lawn-covered slope. There I come face-to-face with an obviously planted row of sequoias. Set into the ground as seedlings in the late nineteenth century not long after the hotel went up, the four trees now tower well over one hundred feet each. At ground level the largest approach five feet in diameter. These imposing trees separate the main hotel building from an 1894 gingerbread satellite structure known as the Moore Cottage. Looking to my right, I see several other wooden buildings offering Victorian-era accommodations for modern travelers.

The very names this place took in its early years reflect the strong bond between the hotel and the Big Trees. When the hotel was built, it was first called "Big Trees Station," but within a few seasons it found a new and more distinctive title: "Wawona." The name is reputed to be a Native American word for the giant sequoias, even though it was bestowed upon the hotel by a woman who made no claim to native ancestry.[1] Regardless, there can be no doubt but that the hotel owes its existence and enduring fame to the nearby Mariposa Grove. Even today, there are those for whom the giant sequoias of the Mariposa Grove and the Wawona Hotel form an inseparable whole. Let us explore how this most durable of giant sequoia resorts came to be.

The 1864 legislation that set aside the Mariposa Grove—withdrawing four square miles of land from the unmanaged public domain and transferring them to the State of California—contained only the most rudimentary instructions about what should be done next. About all the bill did specify was that the grove be managed "for public use and recreation" and that it be overseen by an eight-member commission to be appointed by the governor of California.[2] Implied, of course, was that the public use and recreation would focus on the giant sequoias growing there. Considering the times, California responded promptly, and by the end of September, Governor Frederick Low had issued an interim proclamation of acceptance and appointed the required commissioners. Heading the board was Frederick Law Olmsted, still overseeing the neighboring Rancho Las Mariposas land grant. Most of the other commissioners were either men of note, like California state geologist Josiah Whitney, or men of influence, like Israel Ward Raymond, whose letter to Senator Conness had initiated the move toward legislation. One name stood out, however, as representative of a different sort of Californian: Galen Clark.[3]

Clark had claimed land and set up housekeeping along the banks of the South Fork of the Merced River in 1856 and then been one of the re-discoverers of the Mariposa Grove the following year. At his ranch, as we have already noted, he had early begun to develop a tourist enterprise focused on the Big Trees. As documented by his biographer, Shirley Sargent, Clark's life to that point had not been one of significant accomplishment, but in the years following his personal discovery of the Big Trees, he had demonstrated both his deep interest in them and his willingness to fight for their protection. (Working with Olmsted, Clark had actually spent much of the summer of 1864 in the grove attempting to control the spread of a fire burning thorough the area.) The Yosemite Commission recognized all of this in the spring of 1866, when its members appointed Clark as the first official guardian of both Yosemite Valley and the Mariposa Grove.[4]

From the beginning of his tenure as guardian, Clark was torn between the management needs of nearby Yosemite Valley and his love for the Mariposa Grove. Inevitably, most of his attention was directed toward the valley, where contentious battles had already begun over the fate of land claims filed before the tract was granted to the state. Still, Clark found time to guide traveling parties to his beloved Big Trees, and by necessity, he also continued to be an innkeeper, a calling that remained his primary source of income during these years. To raise money for improvements at his hotel, Clark took on a partner, Edwin Moore, and together, by mortgaging their property on the South Fork, the partnership brought together the $12,000 required to build a wagon road over Chowchilla Mountain to Clark's Station, an effort completed in July 1870.[5] The following season, Clark guided philosopher Ralph Waldo Emerson through the grove. The guardian had become a Yosemite figure of some renown.

Although he was doing good work and had earned the respect of nearly all he met, Clark was spreading himself too thin, and he knew it. In late 1874 he sold his interest in the properties at Clark's Station for little more than the value of debts owed, and in 1880 his first tenure as Yosemite's guardian ended after major changes in the membership of the board of commissioners.[6] Clark's connection with the Mariposa Grove did not end there, but it did change. Although no longer the grove's official guardian, he remained its best-known and most respected guide, a status he would retain into the early years of the twentieth century.

Clark had sold his holdings at Clark's Station to a partnership headed by Henry Washburn, and it was Washburn who now assumed the lead role of managing both the growing resort along the South Fork of the Merced River and its relationship to the nearby Mariposa Grove. Henry Washburn had capital to invest, and after he took control, improvements followed quickly. By July 1875, a wagon road led from Clark's Station to Yosemite Valley, greatly easing the challenge of visiting both that famous place and the Mariposa Grove during the same mountain excursion. When most of the simple hotel complex erected by Clark and Moore burned in November 1878, Washburn immediately initiated work on a grander hotel to replace the lost building. The new hostelry, two stories high and 140 feet long, opened for business on April 1, 1879; he called it "Big Trees Station," but we know it today as the Wawona Hotel.[7] In the months that followed, Washburn also financed a wagon road leading from the new hotel into the Big Trees. The route climbed up through the lower portion of the grove past the Grizzly Giant Tree and then made a loop through the spectacular basin that sheltered the upper part of the grove. One of the features along the new road was Galen Clark's small log-cabin "hospice," first erected for use by horseback touring parties in 1864.

By 1880, in the aftermath of the 1879 hotel improvements and the construction of the Mariposa Grove stage road, the Big Trees near Yosemite Valley were ready to become one of California's premier tourist destinations. It was Washburn's wife, Jean Bruce Washburn, who came up with the next marketing inspiration. In 1882, she began calling the hotel complex "Wawona," a word she suggested came from a Native American word—"wah-wo-nah"—meaning "big tree."[8] This conscious romanticism

Henry Washburn opened a two-story hotel near the Mariposa Grove in the spring of 1879, but it was his wife, Jean Bruce Washburn, who came up with the name that would endure: the Wawona Hotel. COURTESY OF YOSEMITE NPS LIBRARY

had, of course, little to do with the actual native peoples of the area or their ways of thinking about the sequoias as a part of the spirit world that included not only humanity but all of the Sierra's many other life-forms. Instead, the new name seemed intended to provide eager tourists with a destination that was not only unique but also tied to the exotic antiquity of earlier times. This made visiting the sequoias all that much more desirable.

The relative proximity of the Mariposa Grove to Yosemite Valley ensured that word of the Big Trees spread along with stories and images of the famous nearby gorge. As we have seen, photography of both features played a significant role in the creation of the Yosemite Grant in 1864, and Yosemite Valley and the Mariposa Grove continued to receive increasing attention during the following decade. J. M. Hutchings had already written up both places in his pioneering touring book *Scenes of Wonder and Curiosity in California*, published in 1862. Describing the Mariposa Grove, he explained how to get there and detailed the names and dimensions of the larger and most spectacular trees.[9] In 1869, Josiah Whitney brought out *The Yosemite Book: A Description of the Yosemite Valley and the Adjacent Regions of the Sierra Nevada and of the Big Trees of California*.[10] Issued under the auspices of the California Geological Survey, an endeavor headed by Whitney himself, *The Yosemite Book* provided the most complete overview of the Big Trees to date. Whitney summarized what was known at the time about the distribution of the trees, giving primary emphasis to the two major known groves: Calaveras and Mariposa.[11] He listed the biggest trees in both groves, then went on to remark that the trees of Mariposa were notably larger than those at Calaveras despite the ravages of past fires. The first edition was illustrated with photographs taken by Carleton Watkins. Watkins had produced these images during the summers of 1865 and 1866 during visits sponsored by the California Geological Survey. These later trips allowed him to rephotograph many of the scenes he had first captured during his pioneering work there in 1861, including images of the Grizzly Giant and Galen Clark at the base of that same tree.

Art historians have explored the differences in these images, taken some five years apart, and some have interpreted the later set as symbolizing the trauma faced by the Union during the Civil War. From this perspective, the 1861 images emphasize the threat to the nation, as exemplified by the highly visible fire scars on the trees, while the later photographs—the ones in which the fire scars are de-emphasized—document the survival of the nation.[12]

Let us, however, focus on another and perhaps more tangible contrast: that between the two portraits of Galen Clark. In both photos he stands at the base of the Grizzly Giant. The 1861 version presents him clothed in buckskin—a genuine frontiersman. Five years later, Clark has been civilized. In the 1866 photograph, Clark, although still holding his flintlock rifle, now wears a woven wool coat and vest with a checked shirt showing at the collar.[13] Tourism, and the civilized world that came with it, had begun to infiltrate the Yosemite region.

Watkins's work in the area is significant and memorable, but he is just one among many writers, photographers, and painters who produced work related to Yosemite Valley and the Mariposa Grove during the 1860s and 1870s. Here we will follow one particular trail: that of Albert Bierstadt.

By the early 1870s Bierstadt stood as one of the premier landscape painters in North America, an artist as famous in Europe as in the United States, and one whose paintings sold for what would be considered large sums even today.[14] A leading practitioner of the Hudson River School of scenic painting, Bierstadt had early established his reputation for western landscapes. He had first traveled from New York to the Rocky Mountains in 1859, and he made additional trips out West in 1863 and 1871. During these later trips, which included visits to Yosemite, Bierstadt took "notes" for his larger studio paintings by producing oil sketches *en plein aire*. During his 1871–1873 trip, these sketches included many tree portraits from Yosemite and the surrounding Sierra Nevada. Out of these many field sketches ultimately came two grand paintings of the giant sequoias of the Sierra Nevada: *Giant Redwood Trees of California* (1874) and *The Great Trees, Mariposa Grove, California* (1876).

The two works reflect very different perspectives. The 1874 painting features a mountain stream cascading down a forest slope beneath massive sequoia trunks. Native American women—tiny figures near the base of the closest tree—add scale. What defines the image, however, is the rich, golden glow that fills the forest. Even the Big Trees are more golden than reddish. In this image Bierstadt presents us with classic Hudson River School luminism, just the sort of nonrealistic painting that had made his reputation.

The 1876 painting, however, takes a profoundly different approach. Really a portrait of the already famous Grizzly Giant Tree, *The Great Trees* follows its subject upward from its shadowy base to its highest sunlit branches. Unlike Bierstadt's 1874 painting, this work utilizes a more varied palette of colors. A blue Sierra sky contrasts with green foliage; these are much more the colors of the real world, and Lori Vermaas, writing about giant sequoias in 2003 as an art historian, proposed that *The Great Trees* was inspired by Watkins's photographs of the Grizzly Giant.[15] Today, this huge canvas (five by ten feet) both defines how nineteenth-century Americans came to perceive the Big Trees and endures as the best known of all giant-sequoia paintings.[16]

Bierstadt obviously had high hopes for this 1876 painting, for it was one of six that he submitted for display and competition at the Centennial International Exposition that year in Philadelphia.[17] There, to his surprise and disappointment, however, the paintings were not well received. A review of *The Great Trees* in the *New York Times* complained that "the immense upright picture" lacked "any artistic qualities."[18] In truth, although he was only forty-six years old, Bierstadt's popular moment had already passed. Painting styles were moving on, and his grand scenic views no longer enthused the public, which was now leaning toward new trends of painting developing in France. Perhaps the work also failed to impress East Coast art critics because the giant sequoias were no longer the romantic mysteries they had been just a few decades earlier; by now they were part of popular culture—objects of common tourism.*

By 1880, perceptions of the giant sequoias had evolved in significant ways. Through exhibitions, paintings, photography, and popular publications, the Big Trees had achieved unprecedented fame. At the same time, the center of Big Tree tourism had shifted from that first magical grove at Calaveras to the Yosemite region's Mariposa Grove and its new hotel at Big Trees Station (soon to be renamed Wawona). The decline of Calaveras became fully apparent when, after a prolonged dispute over the ownership of the grove and its hotel, both were sold in 1878 on the steps of the Calaveras County courthouse for a mere $15,000. Calaveras's time as the foremost grove of giant sequoias had come to an end. The Mariposa Grove, associated closely with Yosemite Valley, now reigned supreme—an arboreal mecca to be sought out by all who wanted to experience the world's ultimate trees.

---

* It would take more than a century for *The Great Trees, Mariposa Grove, California* to achieve its current status as the ultimate nineteenth-century Big Tree painting.

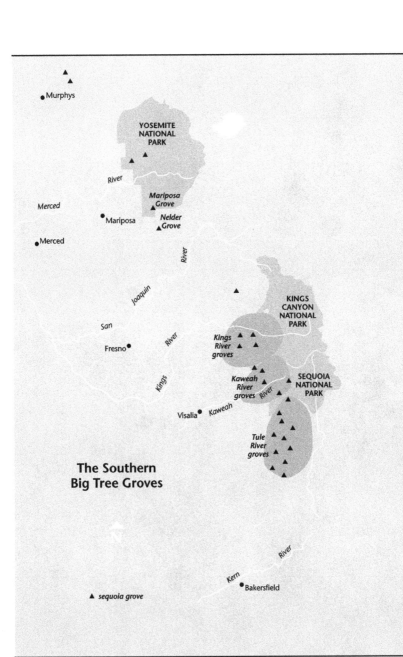

CHAPTER FIVE
Yet Grander Forests

There's a powerful story here, but few know it. To find the site, I need to walk a mile down a quiet trail. The route starts along what obviously was once a road, and, if the sign is to be believed, the primary feature here is an overlook of a reservoir surrounded by YMCA camps. Continuing on, I come to the loop trail's other—and to my mind more significant—feature: a massive upright giant sequoia snag known as the "Dead Giant."

A moment with the topographic map confirms I am narrowly within the boundaries of Kings Canyon National Park, yet it is clear that this tree was not protected when it needed it. The tree's trunk, though time-weathered, confirms that this was once a fine specimen. Measurements give it a ground perimeter of almost ninety feet and a diameter above the basal root swell of twenty-two feet—impressive figures considering that these sums don't include the long-lost bark and sapwood that once clothed the trunk.

I look more closely. The complete absence of both bark and sapwood tells me this tree has been dead a long time, a fact confirmed by the abundant slow-growing lichens that cling to its storm-bleached surface. It is also not hard to figure out what killed it. The marks of the axe that girdled the tree long ago still show. Like the Mother of the Forest at Calaveras, humans killed this tree. But why? I look around for an answer but find none. The National Park Service has not provided an interpretive display here to answer my questions, and when I return to civilization and check the park's website, it offers nothing of substance either. The story is known, however, even if it is not often told.

The Euro-American "discoveries" of giant sequoia trees in the Sierra Nevada did not end with Galen Clark's encounter with the Mariposa Grove in 1857. In fact, the story had barely begun. What would become apparent over the next two decades was that the great majority of the Sierra's Big Trees were in the southern part of the range. Discoveries there came more slowly, however, because those groves were not near the gold rush population centers that had developed in the north in the early 1850s. By the late 1850s, however, unsuccessful miners and entrepreneurs of all sorts were fanning out across California, looking for places where they might make a living or even find a different way to strike it rich.

In the early 1850s, the demand for foodstuffs in the Mother Lode region sent eager ranchers and farmers into parts of California that had not previously been occupied by Euro-Americans. Once there, they usually displaced the native peoples and converted the landscapes for their own needs. One such region was the Kaweah River Delta, a rich, well-watered area of oak forest and green meadows located in the southern portion of California's Great Central Valley. There, the town of Visalia began to take shape, and as that community grew, its residents began to look to the nearby mountains for softwood timber. In their search, they found sequoias. By 1857, ranchers seeking lumber had pushed a rough road up to a mountain valley they called Mill Flat, located on the lower edge of the mountain pine forest along the foothill divide separating the drainages of the Kings and Kaweah Rivers. At that point, it did not take long for these new residents to realize they were less than two miles from an extensive grove of giant sequoia trees. Following the model established at the Calaveras and Mariposa Groves, this latest grove also took the name of the county in which it was located and became known as the Tulare Grove.[1] In 1867, a Visalia woman named Lucretia Baker named the biggest tree in the grove after Civil War general Ulysses S. Grant, who was getting ready to run for president.[2]

Farther south and directly east of Visalia, a cattleman by the name of Hale Tharp followed Native American guides into the great sequoia forest between the Middle and Marble Forks of the Kaweah River in the summer of 1858.[3] This is the sequoia grove we now know as the Giant Forest. Farther south yet, in the Tule River watershed, mentions of "redwood timber" began to crop up as early as 1860.[4] Because of their locations far from the well-populated mining regions and their supply cities, however, these remote discoveries generated little excitement. Yet, these new stands of giant trees were becoming known, and in his *Yosemite Book* of 1869, Josiah Whitney noted: "Here, between the King's [sic] and Kaweah Rivers, is by far the most extensive collection of trees of this species which has yet been discovered in the State."[5] Beyond this general acknowledgment of their existence, however, the full extent of the southern forests remained a mystery to most of the world.

This did not mean, however, that the trees escaped the attention of those who sought to turn them to personal advantage. Among those seeking such an opportunity were two men, William Snediker and William Stegman, who in 1870 were residents of the Mariposa County mining town of Hornitos. Stegman had recently visited the Fresno Grove and brought back a thick slab of bark that had generated considerable interest while on display at the Capitol Saloon in Mariposa. Snediker, meanwhile, had made a trip to the Tulare Grove and come back quite full of ideas. Comparing notes, the two men hatched yet another plan to fell a large sequoia and display it for profit.[6]

Word spread of their intention and, in contrast to reactions a decade earlier at Calaveras, resistance quickly arose. As one Visalia newspaper, the *Tulare Times*, editorialized in its May 21, 1870, edition:

> It Must Not Be Done—It is stated that parties contemplate cutting down one of our Big Trees with a view to removing a section of it to the Eastern States for exhibition. This must not be allowed. These trees stood there thousands of years before Adam's time, and being to all appearances young and thrifty will remain and continue to grow if not disturbed by man. To cut one would be a sacrilege....If the law will not protect them, it should be so amended so that it will.[7]

That same issue of the *Tulare Times* also contained an "instruction" from Edward W. Willett, registrar for the General Land Office in Visalia. Willett's notice moved quickly to its key point: "Persons who cut, destroy, and set fire to the public timber do so at their peril and will be held responsible before the Grand Jury of the United States."[8] Willett's warning was aimed specifically at the project already under way in the Tulare Grove of the Big Trees by Snediker and Stegman.[9] There, on the slopes above Mill Flat, the two entrepreneurs had picked out the first large sequoia they saw and begun to prepare it for destruction and removal. They had already completed the first step, which was to girdle the tree. Willett's notice apparently spooked them, however, and they withdrew, leaving the tree standing but fatally girdled.[10] Known today as the Dead Giant of Grant Grove, it stands there yet.

Snediker and Stegman were not ready to abandon their project, however, so they sought a new location from which to extract a tree. Moving outside the jurisdiction of the General Land Office's Visalia District, they turned their attention to the Fresno Grove, a dozen miles south of the Mariposa Big Trees.* By the end of June, Snediker had succeeded in felling a tree there twenty-four feet in diameter, the largest yet brought down. Not having the tools required to saw through the tree's bole, Snediker instead had his crew dig out and cut the tree's primary roots. This crude approach worked, and the tree came crashing down. Once the huge mass was on the ground, workers followed a familiar technique: hollowing out the lower trunk to form a display chamber. By mid-October, after a 150-mile wagon trip, the "Forest King" arrived at the river-port city of Stockton, where it was displayed for a few weeks before moving on to San Francisco.[11] Shortly thereafter, Snediker and Stegman sold the Forest King to new exhibitors, who, taking advantage of the recently opened transcontinental railroad, moved it east, where it was displayed in Chicago, Cincinnati, and New York. By 1874, the tree had been sold again and set up as a garden folly on the estate of magazine publisher Frank Leslie in Saratoga Springs, New York.[12]

---

* Today this site is known as the Nelder Grove.

Most modern readers will be angered by the quick reduction of this huge tree to, first, a traveling carny attraction and then a lawn ornament for a celebrity. But this tale has a twist. Superficially, this story resembles that of the Mother of the Forest, removed in pieces fifteen years earlier from Calaveras, and yet there are differences worth noting. One detail that stands out is that the excitement generated by the first sequoia exhibits to go east had largely evaporated by the 1870s; the Forest King was not part of an international exhibition but was moved about much like a carnival show before it ended up as a mere garden decoration. Even more significant, though, is how Californians responded to the destruction of the tree. Opponents of its removal sought the help of the government at several levels, and with the aid of the General Land Office staff in Visalia, they prevented the felling of the tree Snediker and Stegman had chosen in the Tulare Grove, even though the effort came too late to prevent the tree's subsequent death.

More significant yet is what came next. It took several years, but in February 1874, John W. Ferguson, a state assemblyman from Fresno County, introduced a bill in the legislature to outlaw the cutting of Big Trees in Fresno, Tulare, and Kern Counties. After several quick rounds of editing, the bill passed both houses and went to the governor for signature in early March. In its final form, the law prohibited the cutting, purposeful burning, or removal of bark from any tree "over sixteen feet in diameter" in the three counties.[13]

Sadly, this idealistic piece of legislation soon revealed itself to be largely ineffective. Barely a year after the passage of the law, a proposal surfaced to remove a tree from the Tulare Grove, this time for national exhibition in Philadelphia in honor of the upcoming centennial of the Declaration of Independence. As the snow melted in the spring of 1875, one Martin Vivian made plans to fell a sequoia and take it east for the festivities. Being cognizant of the state law forbidding such things, he first went to the courthouse in the newly founded railroad town of Fresno, pleaded guilty to cutting a Big Tree, and then paid the maximum designated fine of $300. This done, and now protected from further prosecution by the principle of double jeopardy, Vivian proceeded into the mountains to fell his tree.

Curiously, the reaction to this series of events differed dramatically from the fuss that had erupted when Snediker and Stegman had proposed cutting a tree in the same area five years previously. Reflecting, we might assume, a patriotic appreciation of the nation's centennial, the *Tulare Times*—the same newspaper that had worked to stop tree cutting in 1870—now merely editorialized gently that it might be better to *not* represent Tulare County in this way, but otherwise it did little to stop the project.

Vivian selected for his target the "Daniel Webster," an impressive specimen growing barely one hundred yards from the already famous General Grant Tree. In mid-August, when thirty-seven-year-old John Muir passed through on his first quick visit to the area, he reported meeting "a group of busy men engaged in preparing a butt section of a giant sequoia they had felled for exhibition at the Quaker Centennial."[14] The wandering naturalist was not impressed. In a letter he sent to a San Francisco newspaper he fumed:

> Many a poor, defrauded town dweller will pay his dollar and peep, and gain some dead arithmetical notion of the bigness of our Big Trees, but a true and living knowledge of these tree gods is not to be had at so cheap a rate. As well try to send a section of the storms on which they feed.[15]

What Muir may have said on-site to the men cutting the trees has been lost to time, and regardless, they were not deterred. By September the collected tree materials had been brought down to the railroad that now ran through the Central Valley, and November saw the remains of the Daniel Webster Tree arrive in St. Louis, where they were set up and opened to the public. The following spring, the exhibit made it to Philadelphia, where it failed to gain admission to the official centennial exhibition. Instead, it was erected in the informal "sideshow" district across the street from the exhibition's main entrance, where it remained on display until September 9, 1876, when it was consumed by a fire that destroyed the buildings housing the sideshow attractions. Muir, if he read about the whole sorry mess, surely shook his head in disgust.*

---

* The "Centennial Stump" and the fallen log that forms the mortal remains of the Daniel Webster Tree can still be seen near the General Grant Tree today.

If the stories presented here seem contradictory, we must admit that they are. The discovery of extensive groves of giant sequoias south of the Kings River brought out both the best and the worst attitudes about the Big Trees. The same groups that thwarted the 1870 effort to cut an exhibit tree at the Tulare Grove in 1870 acquiesced five years later to an even more egregious felling in the same area. (I say "more egregious" because the tree felled by Vivian was adjacent to the already famous General Grant Tree and thus plainly visible to visitors.) Yet this same confusing saga also produced a law plainly intended to protect the Big Trees from exploitation and destruction. That this legislation would have no practical effect does not completely obviate its significance. Even as the public had begun to see sequoia exhibits as no more than carnival sideshows, efforts to protect standing trees were gaining adherents. First there had been the Yosemite Grant. Now, at least in theory, none of the largest trees in the southern Sierra were to be cut either. Things would not turn out to be that simple, of course, but popular ideas about the Big Trees had begun to move in new, if sometimes contradictory, directions.

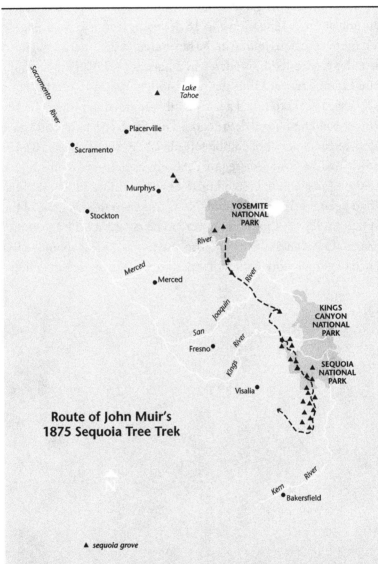

CHAPTER SIX

A Wandering Scot

The fire-hollowed sequoia snag stands in a second-growth fir thicket a few dozen yards below a logging road. No sign marks this ancient arboreal ruin, which must once have been one of the very largest sequoias. To find the hulking remains, one must simply stay alert while driving along the quiet dirt road. I pull over and park at a wide spot some fifty yards past where I glimpsed the snag. A faint trace of a path leads down the hill toward the towering black mass. I stumble down the slope and step over several foot-thick fallen trees. Few visit this place.

I inspect the massive object. A somber black column of fire-scarred wood rises into the forest canopy. Repeated fires have burned so deeply into one side of the dead tree's base that its growth rings are exposed almost to the center. Old hatchet marks document long-ago attempts to clean these rings of charcoal and lichen so they could be counted. A careful inspection of the burned surface also discloses a few names and dates inscribed into the heartwood. All are at least a century old.

This tree once enjoyed a brief moment of literary prominence, yet for much of the twentieth century, no one knew exactly where it stood. The tree first gained attention in an essay published in San Francisco's *Daily Evening*

Bulletin on October 22, 1875. The correspondent, talking about the continuing effort to find the biggest trees, wrote: "The largest measured by me is a stump 80 or 90 feet high, situated on the south side of the Middle Fork of the King's [*sic*] River. At a height of 4 feet above the ground it is 35 feet 8 inches inside the bark, and a plank this wide could be obtained from it of solid wood, without a decayed fiber."[1] In a follow-up article that came out in *Harper's New Monthly Magazine* in November 1878, the same writer returned to the tree, again discussing its dimensions and adding the provocative fact that it "is probably over four thousand years old."[2] These pronouncements marked the arrival of a new student of the Big Trees: John Muir. Over the next four decades, he would become the sequoia's best-known public advocate and devotee. In the autumn of 1875, however, he was only just beginning.

The Big Trees of the Sierra Nevada fascinated nineteenth-century botanists every bit as much as they did the tourists. We have seen already how the British botanical community latched onto the trees almost from the moment their presence became known, an interest that manifested itself in both a scientific name for the species as well as the tree's broad distribution for horticultural purposes. The commercial exhibitors of the trees also made attempts to interpret them as biological marvels, but much of what they disseminated fell more into the realm of myth than genuine scientific description. None of this should surprise us. In the biological world, the middle years of the nineteenth century were the apex of the great age of taxonomy. For the world's botanists—as well as their friends in the zoological and ornithological fields—the primary goal was to find out what lived on our planet and where. Just to know that the trees existed was enough for most, but there were some who began to ask more complex questions beyond merely cataloging and collecting.

Among these curious souls was a thirty-four-year-old Scottish immigrant scratching out a thin living in and about Yosemite Valley. John Muir had arrived at the famous glacial gorge in 1868 only weeks after setting foot in California. On that first, short visit to the higher reaches of the Sierra, Muir also visited the Mariposa Grove. There he made the acquaintance of Galen Clark, who would become a lifelong friend.[3] During the next several years, Muir worked hard to eke out a minimal living, tending sheep in both the foothills and the High Sierra, then working in a sawmill in Yosemite Valley that cut up trees that, according to Muir, had already fallen. His employer in this endeavor was none other than magazine editor J. M. Hutchings, who by now had settled in Yosemite Valley and become an innkeeper. By the early 1870s, Muir had largely escaped these jobs, which he thoroughly detested, and was making a living as a guide for the region's growing stream of tourists. In this he relied to a considerable degree on the connections and recommendations of friends he had made during his several years at the University of Wisconsin in the early 1860s.

It would be misleading to describe Muir in these years as a full-fledged scientist. He loved wildness and all the forms of nature that inhabited wild places, but to him the natural world was as much a source of mystical joy as of scientific knowledge. He found in untrammeled nature a powerful antidote to his Scottish father's stern Calvinism, and Muir's early reactions to the sequoias reflect his own enthusiastic, joyous approach to nature. In an undated letter apparently written in the very early 1870s, Muir waxes rhapsodic about what he called "King Sequoia":

> The King tree and I have sworn eternal love—sworn it without swearing, and I've taken the sacrament with Douglas squirrel, drunk Sequoia wine, Sequoia blood, and with its rosy purple drops I am writing this woody gospel letter.[4]

Muir made his "sequoia wine" by dissolving sequoia cone pigment—a crystalline substance found in the cones—in water. The result was a bitter-tasting reddish liquid much the same color as the trunks of the Big Trees themselves. It is hard to miss the religious symbolism here.

At the same time that he reveled in the mystical side of the trees, Muir also had a naturalist's strong urge to understand what he was seeing. His greatest interest during these years was to discern the geological origins of Yosemite Valley, but during this effort, he also continued to seek out and appreciate the Sierra's widely diverse plant life. He had been a collector of plant specimens as early as his days at the University of Wisconsin, and now, to indulge his interests as well as to strengthen his personal connections in the larger world of natural history, Muir began to collect and share specimens from the Sierra.[5] Among those to whom Muir sent plant samples was Asa Gray of Harvard University.

By the early 1870s, Gray stood at the very apex of the American botanical community. A professor of natural history at Harvard since 1842, Gray had built over the years a formidable network of botanical collectors, whose purpose was to fuel the growth of taxonomic knowledge as well as the herbarium at Harvard. Gray himself had learned much of his botany from John Torrey, with whom he worked on several long-term projects, including a never-completed effort to publish a multivolume flora containing all the plants of North America.

Among Gray's many connections were Ezra and Jeanne Carr. Professor Carr and his wife, a botanist in her own right, had befriended Muir during his university days in Madison in the early 1860s, and Muir and Mrs. Carr remained close through a prolonged correspondence. In 1869, the Carrs moved west when the professor accepted a position at the fledgling University of California, and Muir became even closer to the pair.[6] It was through this network of acquaintances that Muir began to correspond with Asa Gray.

In the summer of 1872, Muir's relationship with Gray took a major step forward when Gray embarked upon a long-dreamt-of first journey to the Far West. That July, Gray visited both the Calaveras Grove and Yosemite, where he sought out Muir. Together, they toured and collected wild plant specimens. Although they had enjoyed a productive correspondence, face to face the two men found their personalities not particularly compatible. Muir, ever the exuberant soul, remarked later about Gray's "angular factiness."[7] Nonetheless, the two found much of mutual interest during their ten days together, a period that included an obligatory visit to the Mariposa Grove to see the Big Trees.

Like many before and after, Gray was fascinated by the Big Trees. He remarked upon their "grandeur" and their "singular majesty" and lamented the all-too-human urge to name them after contemporary people.[8] But bigger questions soon engaged him, including their seeming isolation, both geographically and biologically. Gray was an early and outspoken supporter of Charles Darwin's theory of evolution, and as he studied the sequoias he found much to think about from that perspective. As Gray returned east on the new transcontinental railroad, these thoughts matured, and by the time he arrived in Dubuque, Iowa, for the annual meeting of the American Association for the Advancement of Science, his ideas had come together. As outgoing president of the association, Gray was expected to deliver an address reflecting his own scientific inquiries. Inspired by what he had seen in the Far West, Gray titled his speech "Sequoia and Its History."

As a botanist with a robust commitment to evolution, Gray's inquiry into the history of the giant sequoias focused on biological questions with clear evolutionary implications. "Have these famous Sequoias played in former times and upon a larger stage a more imposing part, of which the present is but an epilogue?" he asked.[9] "Or are they now coming upon the stage (or rather were they coming but for man's interference) to play a part in the future?"[10]

Using these questions as an opening, Gray marched off into a commanding defense of evolution, his presentation all the more powerful for the fact that he wrote it while traveling, presumably without access to his library resources at Harvard. In his speech, Gray explored the differences between the climates and the resulting natural vegetation of the two coasts of North America, then went on to compare these regions to eastern Asia and Europe and the forests found there. He connected the two sequoia species of California (the California coastal redwood was swept in with the Big Trees in his analysis) with the bald cypress (*Taxodium distichum*) of the American South and the *Glyptostrobus* trees of China.[11] He went on to relate these trees to the distribution of living specimens as well as fossils found in North America, Europe, and the Arctic. In these fossils he identified a number of extinct forms that he considered members of, or at least ancestors of, the genus *Sequoia*, including trees that once grew in Alaska, northern Canada, Greenland, the Spitsbergen area of Norway, and northern Europe. In conclusion, he found the two *Sequoia* species of California to be the living descendants of an ancient and once widespread lineage that had since been isolated in California by patterns of climate change during and after the Pleistocene ice age.[12]

Gray's entire talk, running to some twenty pages of type, was published in the October 1872 edition of the *American Naturalist*.[13] Certainly, one of the more avid readers of the essay would have been John Muir. By the winter of 1872/73, the Californian was well along on his personal campaign to become not only a naturalist and guide but also a writer. In late 1871, he had published his first paid newspaper piece, an essay titled "Yosemite Glaciers" that ran in the *New York Tribune* on December 5.[14] The following summer he sold several more articles, including two to the *Overland Monthly* magazine. During 1873, he marketed several additional natural history essays, but increasingly his attention was focused on geology. In the fall of 1871 he had discovered the Lyell Glacier, the first to be identified in California, and this had strengthened his belief that the landscapes of the Sierra Nevada, especially Yosemite Valley, had been profoundly shaped by glacial ice. As a writer, he brought it all together in the summer and fall of 1874 in a series of six *Overland Monthly* essays running under the series title "Studies in the Sierra."[15] Muir compiled a summary of key points from these essays and submitted it to the American Association for the Advancement of Science. His friend Asa Gray, now the association's former president, read the paper at the society's August meeting in Hartford, Connecticut, and saw that it was printed in the organization's *Proceedings* several months later.[16]

Muir now began working on another series of newspaper articles, this one titled "Summering in the Sierra," but he continued to ponder the questions Gray had asked about the sequoia in his 1872 talk. In "Sequoia and Its History," Gray had written about both the lack of information regarding the southern part of the giant sequoia's range and the seeming paucity of natural reproduction in the groves, a circumstance that Gray implied threatened the future of the species. Here, Muir knew, were questions to which he could apply his unique ability to wander and write. In late August 1875, Muir set out from Yosemite to explore the natural world of giant sequoias.

The journey, significant in a number of ways, would be Muir's last great wilderness trip as a solitary walker. In coming years, as his reputation increased, his adventures would more often occur in the company of others, as Muir filled the roles of teacher, guide, and prophet. But in the autumn of 1875, except for a tough mule named Brownie, Muir wandered south alone. He kept a journal as he explored, and as he traveled he wrote a number of newspaper articles that he sent to the *Daily Evening Bulletin* in San Francisco. In the coming years, Muir would rework this material repeatedly, turning various pieces into a scientific paper, an article for a national magazine, and, ultimately, chapters for several of his books. As the years passed, the stories grew richer and the language more flowery, perhaps a testament to the fact that Muir's affection for the original adventure never faded.[17]

Heading south on foot from the well-known sequoias of the Mariposa Grove, Muir first explored the Fresno (now Nelder) Grove, located hardly more than a day's walk away. His correspondent's report from there, published on September 21, tells us much about what was on his mind as he began the trip.[18] Moving beyond the hyperbole that had characterized so much early writing about sequoias (and still often does), Muir approached the Fresno Grove by doing what he did best: studying it as an observant naturalist. His first letter sketched a detailed and accurate picture of the trees and their forest setting. He explored the color of their wood, the nature of the cones and seeds, and the durability of fallen logs. Only after he had provided his readers with this substantial natural history did he finally relay a short human-interest story about forest hermit John A. Nelder, whom he found living in a small cabin in the grove.

Over the next two weeks, Muir moved farther south, surveying the mid-altitude forest areas of both the small Fresno River watershed and the immensely larger San Joaquin drainage. To his surprise, he found no sequoias there. In fact, the next grove did not appear until he had traveled nearly forty air miles to the southeast, a slow journey that included crossing the rugged canyon of the San Joaquin River and climbing over the divide that separated the San Joaquin from the watershed of the Kings River. Pausing only briefly to take the relatively small grove's measure, Muir moved on, driven by the knowledge that winter in the mountains could not be too far off.[19]

Challenging Muir on this journey was the fact that he was traveling "across the grain" of some of the roughest terrain in North America. By heading south through the mountains and attempting to stay within the range's mid-altitude forest belt, Muir had to cross multiple canyons and intervening divides. By early October he was working his way down into the largest of all these obstacles, the huge lower gorge of the main stem of the Kings River, a cross-country plunge that took Muir and Brownie downhill a full 7,000 vertical feet and then back up again almost as far. But Muir persevered. On the ridges along the canyon's south rim he could see sequoia crowns emerging from the thick forest that clothed the slopes.

What came next justified Muir's entire trip. For the next several weeks, as autumn faded toward winter, Muir wandered through an almost continuous forest of giant sequoias. The existence of at least portions of this forest had been known for a decade or longer, but no one had yet traveled north–south through the region with the specific goal of mapping out the extent of this great forest of Big Trees. Muir shared a first glimpse into this new world in an article published in the *Daily Evening Bulletin* on October 22, a full month after his report from the Fresno Grove. In "The Giant Forests of the Kaweah," the careful reader can piece together Muir's general route as it took him up out of the Kings River canyon to Mill Flat, where he found Thomas's Mill at work cutting away at the surrounding forest.[20] Northeast of Mill Flat, he visited the great sequoia forest of Converse Basin, the largest by far he had ever seen. There, while exploring, he found the huge burned-out hulk of the dead sequoia tree that impressed him so. In this snag he recognized not only evidence of the tree's original massive size but also, because of its deep fire scars, a rare opportunity: the potential to examine its annual growth rings all the way to the tree's heart. Muir spent a day chopping away at the burned surface until he could see the rings, and then he counted them. He estimated that the tree was at least 4,000 years old.* He also noted, with alarm, that plans were already under way to log the grove.[21]

Moving on, Muir studied the trees in the already well-known "King's [*sic*] River Grove," the home of the General Grant Tree.[22] This visit came just months after one of the largest trees in the grove—the specimen named for statesman Daniel Webster—had been felled, broken apart, and partially removed so that it could be exhibited the following summer at the Centennial International Exposition in Philadelphia. Muir took advantage of the tree's destruction to collect more data about the age

---

* Today, the US Forest Service estimates the snag's life span as about 3,200 years.

of the sequoias. After talking to workers at the site, he recorded that three different persons had counted the rings and come up with ages ranging from 2,125 to 2,317 years.[23] This is just the sort of true-observation information he had come south to find!

Leaving the Kings River Grove, Muir continued southward into the Kaweah watershed, discovering sequoias growing vigorously across the landscape in all directions. On the western slope of what is now called Redwood Mountain, Muir discovered Hyde's Mill, a small operation busy despoiling the surrounding woods.

> In this glorious forest the mill was busy, forming a sore, sad center of destruction, though small as yet, so immensely heavy was the growth. Only the smaller and most accessible of the trees were being cut. The logs, from three to ten or twelve feet in diameter, were dragged or rolled with long strings of oxen into a chute and sent flying down the steep mountain-side to the mill flat, where the largest of them were blasted into manageable dimensions for the saws. And as the timber is very brash, by this blasting and careless felling on uneven ground, half or three fourths of the timber was wasted.[24]

Muir continued his southward wanderings, and by the time he completed the draft of his second *Daily Evening Bulletin* piece, he had arrived in another very large grove, located on the mountaintop plateau along the northern rim of the great canyon of the Middle Fork of the Kaweah River. Later, Muir would claim to have named this place "The Giant Forest" during this visit, but he made no such declaration in the *Daily Evening Bulletin*.[25] Instead he simply gushed about the trees:

> No amount of familiar communion with the small companies of trees that occur in Calaveras and Tuolumne counties can yield anything more than feeble hints of the sublimity of this grand sequoia realm....In the morning, leaving my jaded mule in the meadow, I sauntered free in this solemn wilderness. Go where I would, sequoia reigned supreme. Trees of every age and size covered the craggiest ridges as well as fertile, deep-soiled slopes, and planted their colossal shafts along every brook and along the margins of spongy bogs and meadows....No forest could equal this.[26]

The next article Muir prepared for the *Daily Evening Bulletin* apparently disappeared in the mail and was never published, but we can piece together his progress from his journal and later descriptive writings. The naturalist worked his way through the sequoia groves of the several forks of the Kaweah and over the divide into the equally impressive forests of the Tule River watershed. He measured trees, spent time watching what was most likely a lightning fire move slowly through a sequoia grove, and continued to feel guilty about what he was doing to his poor mule. By the middle of October he had made his way to the southern limit of the groves at Deer Creek, and then, once he was sure he had reached his goal, he descended into the San Joaquin Valley and walked north toward home. On the ground, he had traversed something like six hundred miles in eight weeks, nearly all of it in unmapped and trail-less terrain.

Although he wrote and published only two relatively short essays during the mountain portion of this trip, Muir's autumn 1875 journey in search of sequoias ultimately would revolutionize public perceptions of the species and its status. Once settled in for the winter, Muir sat down with his trip journal and the other materials he had used to draft the preliminary reports he had rendered for the *Daily Evening Bulletin* and began to make a much fuller accounting of what he had learned about the giant sequoias of the Sierra Nevada. Over the next year, this effort produced both a scientific paper that spoke to Asa Gray's questions as well as an extended piece in a national magazine that detailed the existence of what he called "The New Sequoia Forests of California." Read together, they are highly informative even today.

The scientific paper came out first. Muir had it ready to be presented at the annual meeting of the American Association for the Advancement of Science, which took place in Buffalo in August 1876. Once again, Asa Gray read the paper to the delegates, undoubtedly noting how well it responded to many of the questions he himself had put before the same group back in 1872.[27] Muir positioned his paper as "On the Post-Glacial History of Sequoia Gigantea," but the text covered a good deal more than that. He began by laying out an ambitious list of questions:

What area does Sequoia now occupy as the principal tree? Was the species ever more extensively distributed on the Sierra during post-glacial times? Is the species verging on extinction? And if so, then to what causes will its extinction be due? What have been its relations to climate, to soil, and to other coniferous trees with which it is associated? What are those relations now? What are they likely to be in the future?[28]

In the pages that followed, Muir methodically addressed these questions, working hard to keep his language simple and analytical. He described the tree's range, giving the first full description of the locations and extent of the southern groves. Responding to the question of whether the sequoias had been more widespread in post-glacial times, he concluded that they had not. He based this argument on evidence he had discovered that showed that both dead sequoias and the root craters they left behind in the soil were highly durable and yet completely missing from Sierra forests outside the existing groves.

Moving on from there to the heart of what Gray had speculated about in 1872, Muir explored several questions relating to the possibility of sequoia extinction. Here, he had real news. The southern groves, he explained, were unlike the smaller northern groves in several ways. Not only were they larger, but the sequoias in the southern groves also seemed better adapted to their chosen world. In the south, they dominated the forest in a way they did not up north, and the degree and frequency of reproduction was strikingly higher as well. One could not visit the southern groves, he asserted, without being struck by the fact that the sequoias were prospering there, well suited to the region's climate and soils. This was not a species that was about to fade away on its own, he assured readers. But this led him to another question: If the trees were doing so well in the southern half of their range, what could explain the large geographical gap between the trees south of the Kings River and the northerly Fresno, Mariposa, and Calaveras Groves? Here, Muir connected the dots in a way that modern readers might call interdisciplinary—that is, he used the knowledge he had collected during his glacial studies to point out the likely relationship between the paths of the largest Sierra glaciers and gaps in the range of the sequoias.

At the same time that Muir worked through his journal notes to draft the scientific paper for the American Association for the Advancement of Science, he was also crafting the same materials into a popular narrative suitable for general readers. As a budding writer, Muir had broken through to a national magazine audience for the first time just as he was wrapping up his sequoia wanderings; *Harper's New Monthly Magazine* published his essay "The Living Glaciers of California" in November 1875.[29] Founded in 1850, *Harper's* supplied the quantity and quality of readers to essentially confirm Muir's arrival as a writer of national status. During 1877, the naturalist fed several more articles to *Harper's*, and it was for that organ that Muir worked up his popular essay about his giant sequoia expedition. "The New Sequoia Forests of California" ran in the magazine's November 1878 issue.[30]

Substantially longer than the scientific essay he had sent to Gray, Muir's *Harper's* piece offered its audience a much fuller reading experience. The obvious (and necessary, considering the context) difference was that Muir spent a good deal of time in the essay detailing his personal adventures during the trip. Brownie the mule and hermit John Nelder received considerable attention, as did the many domestic sheep grazing in the Sierra and the shepherds looking after them. But at the same time, Muir also made sure that nearly all of the material he had developed documenting the natural history of the Big Trees and their range found a place in the essay. He also worked in his answers to Gray's questions, each of them still carefully flagged in italics. The entire essay ran to about 8,000 words and provided its national audience with by far the most complete and accurate overview of the Big Trees of California yet to appear in print. Using both his skills as a naturalist and his growing expository expertise, Muir had moved well beyond the shallow fund of semi-mythical stories and exaggerated statistics that had typified most giant sequoia literature to date.

Although Muir had focused his attention in both essays heavily on questions related to the evolutionary outlook for the sequoias, the primary effect of his writing was to bring clearly into the public's mind a new awareness of and appreciation for the large and vigorous groves of giant sequoias to be found in the Sierra Nevada south of the Kings River. Previously, these forests had been little known beyond their immediate environs, and what had been written about them had been mostly speculation. Now, for the first time, both the extent and the majesty of these southern forests had been published and widely distributed, and not just to California newspaper readers but also to scientists and magazine readers nationwide. For the moment, the southern groves remained remote and difficult to visit, but within a dozen years of the publication of "The New Sequoia Forests of California," a major part of the sequoia's southern range would be set aside as the nation's second and fourth national parks. Once discovered, these trees would not be forgotten.

Muir's 1875 wanderings through the realm of the Big Trees also changed Muir himself. The writing he produced after the trip made it clear that he was not only learning how to be a successful writer but was also rethinking his personal relationship with the natural world. He had first gone into the Sierra in 1868 full of simple, unbridled appreciation for the beauties of nature. Now, through both his geological work on glaciers and his botanical efforts related to the Big Trees, he had established himself in scientific circles as an observer and naturalist worth following. But there was more. Muir went south through the sequoia groves pursuing Gray's concern that the trees might be moving naturally toward extinction. Although Muir concluded that this would not likely occur in the foreseeable future, he found something else that did worry him. The threat to the sequoias—and there was one—came not from the inability of the trees to live in the modern era but rather from the actions of contemporary humans. If current trends continued, Muir grasped during his journey, loggers would soon decimate their numbers and thus immeasurably impoverish succeeding generations.

In September and October of 1875, Muir found small sawmills assaulting the sequoias in multiple locations, including on the western edge of the Kings River Grove and the lower fringes of the Tule River sequoia stands. He also collected stories of much bigger plans—plans that would in the next decade begin the ultimate destruction of most of the old-growth sequoias in places like the Fresno and Converse Basin Groves. This realization shocked Muir into a new sort of writing, a form not visible in his earlier work. Reflecting this new concern, Muir's first published piece about the sequoias after his return to civilization in late 1875 was not the two essays described above but rather a short submission addressed to the *Sacramento Record-Union*. Titled "God's First Temples: How Shall We Preserve Our Forests?" the piece was printed on February 5, 1876, barely two months after Muir's great sequoia adventure had ended.[31] The text began with a brief overview of the current distribution and status of the species, and then Muir moved to his primary point:

> But waste and pure destruction are already taking place at a terrible rate, and unless protective measures be speedily invented and enforced, in a few years this noblest tree species in the world will present only a few hacked and scarred remnants.[32]

He went on to excoriate the effect of numerous fires ignited accidentally in the Sierra by careless travelers, as well as the damage done to the vegetation by the huge flocks of domestic sheep that had invaded the Sierra in recent years. The challenge, as he saw it, was for a new kind of thinking that would change both society and government.

Employing the hindsight that comes with time, we can see clearly now the significance of Muir's giant sequoia wanderings in the fall of 1875. During and after that trip, two shifts in thinking occurred that would change the course of California history. Not only did John Muir make the world aware of the location of the best of the sequoias—a key step in their later protection—but also, and at the same time, the naturalist was himself changed. Before he trekked through the southern sequoia groves, Muir was a joyous celebrant of nature. Afterward, in a way that would only grow for the remainder of his long life, Muir moved ever more toward being a public advocate and activist for the preservation of nature.

It is not too much to say that in 1875 the sequoias played a key role in finalizing the mature character of the John Muir we still honor today.

CHAPTER SEVEN

Free for the Taking

I walk down the rutted dirt trail from the picnic ground. As I descend, the brush thickens and the noises of the parking area fade into the background. I begin to hear birdcalls and the buzzing of insects in the willows that line the path. A deer jumps out of its hiding place and trots down the trail, cautious but not panicked. Scattered in the brush I start catching glimpses of giant fragments of weathered wood, and then a massive wooden mountain rises out of the manzanita—a fire-charred stump a dozen feet across at its summit. More stumps appear as I come out of the brush on the edge of a wet meadow. I begin to count; at least a dozen large sequoias fell here to the axe. More stumps, undoubtedly, are hidden in the second-growth woods that surround this wetland. I try to imagine the forest that grew here before the loggers came. That old growth is long gone now, of course, replaced by brush and younger trees. Some of the sequoias that sprouted here in the aftermath of logging have made it well beyond the sapling stage, their spire tops pointing skyward, offering promise that some will eventually grow to fill the niches left vacant by their harvested parents. Still, it will be centuries before monarch sequoias again dominate this battered place.

I pull a trail brochure out of my pocket. I found the pamphlet, a souvenir of my ranger days, on my bookshelf at home, and although it's not the latest

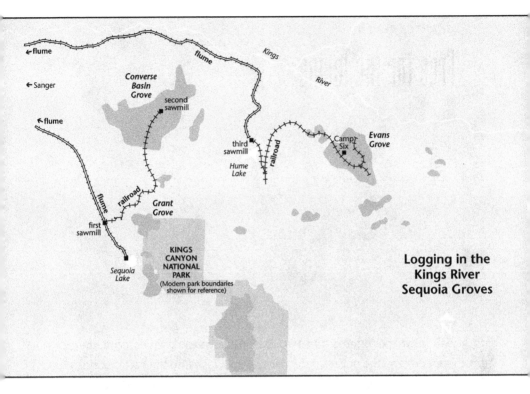

edition, the story it outlines can still be easily followed on the ground. A pioneer entrepreneur named Smith Comstock erected a steam-powered sawmill here in 1883, clearly inspired by the potential of harvesting huge trees. Comstock logged here for six summers before exhausting the site and moving his mill to another location. We call what he left behind "Big Stump Basin," and the irony is that the abused forest is now a preserved and protected feature within Kings Canyon National Park.

My brochure, first issued half a century ago and since reprinted numerous times, urges me to reflect on the destructive powers of humanity, and certainly one can find much at Big Stump to consider. To twenty-first-century minds, accustomed as we are to seeing sequoias as near-sacred objects, their wholesale destruction for profit feels profoundly wrong. Yet the story here is a familiar one in American history. In the latter years of the nineteenth century, even while some approached the trees as artists or scientists, others arrived with sequoia-sized felling tools. Before they were done, decades later, thousands of Big Trees would fall to axes and saws.

The Euro-American culture that arrived in California in the middle years of the nineteenth century had literally cut its way west from the Atlantic seaboard. From the time of the first settlements on this continent in the seventeenth century, Americans had felled trees. The great cut began in the hardwood forests of the Atlantic Coast, crossed the Appalachian Mountains, then worked its way across both the Ohio Valley and the Deep South. Everywhere, trees fell—some to provide lumber, but most just to free the land for farming. By the 1850s, both settlement and tree cutting had crossed the Mississippi and were approaching the edge of the largely tree-free Great Plains.

Meanwhile in the land that would become the state of California, logging had begun with the Spaniards, who cut redwoods along the coast to provide lumber for their colonial settlements of the late eighteenth century. Small-scale timber harvest continued during the Mexican era in California but remained limited to coastal areas. All this changed, of course, with the gold rush and the arrival of its several hundred thousand miners. Gold mining required lumber to build flumes, stabilize underground tunnels, and erect structures. Wood also provided the energy that fueled the mining industry. It should not surprise us that the actual discovery of gold in California in 1848 occurred during the construction of a sawmill in the Sierra foothills by Euro-American entrepreneurs.

The city by the Golden Gate that appeared almost overnight at the beginning of the gold rush was initially built almost entirely of wood, and when it burned—which it did several times during the early 1850s—it was rebuilt of the same materials. The demand for lumber in California soon became almost insatiable.

So, in many ways, it was only a matter of time before Californians began to look at the Big Trees of the Sierra Nevada and see opportunities for logging. Several factors slowed the beginning of what was to come. At first, when the trees were only known to exist in a very few select locales, tourism offered profit enough. Other factors also conspired to delay logging, including the often remote locations of the groves and the fact that many pine forests in the Sierra were much easier to access than sites with sequoias. Equally discouraging was the fact that felling and milling giant sequoias required special skills and equipment; this was not the sort of work successfully undertaken by small groups of men with only hand tools and muscle at their disposal.

The milling of sequoias also required that the logs be moved—no easy feat in the nineteenth century. Modern students of Sierra Nevada history often wonder why early lumbermen bothered to cut the Big Trees at all. Not only were the felled logs enormously difficult to move, but also the brittle nature of the trees' boles often resulted in major parts of the logs shattering into useless fragments when the massive trunks hit the ground. Commonly, more than half of a sequoia log would be lost to breakage. And yet, the attraction of so much clear lumber in so many single trees could not be resisted. There had to be a way to make money felling sequoias. If they could not be reduced to good saw timber (the wood was too brittle and prone to breakage to be used effectively that way), God must have put them in the Sierra for some purpose.

Public reaction to cutting down the trees was certainly not a deterrent. The state law that had been passed in 1874 to prevent logging of the largest giant sequoias in the southern Sierra would never be invoked in the decades that followed. To most Californians, it made sense to set aside a few of the Big Trees for tourist purposes but then cut the rest.

The spread of logging into the giant sequoia areas of the Sierra Nevada also required a local market for the lumber produced. Initially, such a market did not exist beyond the gold rush region where the northern groves were located, but by the early 1870s this situation began to change as the Central Pacific Railroad pushed its line southward down the Central Valley. The railroad reached Tulare County in the summer 1872, and along the line, a string of towns developed, each one requiring lumber. Looking eastward across the largely treeless arid plains of the valley, residents and farmers alike noticed the Sierra Nevada and its extensive softwood forests. Soon, trees began to fall.

Even if many of the sequoia groves were initially protected in a de facto sense by geography, in three regions at least, the terrain allowed early lumbermen to push wagon roads into the Big Tree forests and establish mills. The Fresno (Nelder) Grove represented the northernmost of these areas. During his 1875 visit, Muir collected information about planned logging in the vicinity, and the first of several small mills in the area opened soon thereafter. Logging would continue there on and off for decades, with about half of the grove's large sequoias eventually being felled.¹

To the south, in Tulare County, early sequoia logging focused on two other regions. The groves along the divide separating the Kings and Kaweah Rivers attracted early attention. Lumbering began a few short miles from the western edge of the Kings River groves in the early 1860s with the construction of People's Mill at Mill Flat. This outfit approached the edges of the modern Grant, Sequoia Creek, and Big Stump Groves but did not actually cut much sequoia wood. A few miles to the southeast, however, the story was quite different. There, on the western slope of a ridge still known as Redwood Mountain, Hyde's Mill began cutting Big Trees in 1873. By the time Muir passed through the area in late 1875, the destruction had taken on significant proportions. In his journal Muir fumed:

> Hyde's Mill, booming and moaning like a bad ghost, has destroyed many a fine tree from this wood—two million feet of lumber this year. And it has been running for three years. When felled, the sequoia breaks like glass—from twenty-five to fifty per cent unfit for the mill.²

Hyde's Mill would break ground as the first Sierra Nevada logging outfit to focus primarily on old-growth giant sequoia. Between 1873 and 1877, several hundred trees came down. For practical reasons, crews ignored the largest trees, which were simply too difficult to fell and move using axes and oxen. But trees up to twelve feet in diameter came crashing down and were dragged to the saws.³ More than any other, this was the sawmill that shocked Muir and inspired his emergence as a public advocate for the protection of sequoias over the winter of 1875/76.

A few miles to the south, the extremely rugged terrain surrounding the sequoia groves in the heart of the Kaweah River watershed protected them for the moment, but farther south yet, in the Tule River country, terrain again allowed loggers access to the sequoia groves. In the mid-1870s, limited logging of sequoias began in both the Dillonwood and Mountain Home Groves.⁴ And all of this was simply a small-scale warm-up to what would come in the next decade.

To this point, most sequoia logging outfits had been conducted on a relatively limited basis, restricted not only by the challenges of logging in the rugged Sierra but also by the fact that the national government had yet to create a legal mechanism that allowed lumbermen to gain legal title to large tracts of forest land. In 1878, Congress removed this problem, thus opening the door to a new kind of large-scale giant sequoia logging. The change came in the form of the federal Timber and Stone Act, a statute that expressed America's frontier attitudes toward forest resources by authorizing the government's General Land Office to sell lands that had forest or mineral value but were not suitable for cultivation; they were parceled out in blocks of 160 acres. The law established the minimum value for this land as $2.50 per acre, and almost all Timber and Stone tracts would sell at that price. In theory, the law limited purchases to 160 acres, but the statute had been written so laxly that in practice many speculators used it to accumulate large tracts of forest. The law applied specifically to public lands in California, Oregon, Washington, and Nevada, including the Sierra.

This is what the lumber industry had been waiting for. It was now possible, with only minimal effort, to take control of the Sierra's forests for about $1,600 per square mile. For those who wanted to abuse the law, the process offered great potential. Stories abound of lumber companies hiring men in saloons to go out into the woods and each claim their 160 acres of forest. As soon as the claims were filed, the lumber company would pay off the entry men and legally purchase the claims. Over the next decade, enormous acreages would be sold, much of that land ending up in the hands of large lumber companies.[5]

As a result of the Timber and Stone Act, claims were filed on two of the grandest of all the Sierra's sequoia groves: Converse Basin and the Giant Forest. At the same time, lumbermen also went after the wonderful Big Trees groves located near the headwaters of the Tule River. Each of these stories would turn out differently. Only in the last of these three cases did the law work in a way that reasonably resembled what Congress had intended.

In the journal notes he made during his 1875 journey, John Muir had written that he thought the Tule River forests were perhaps the finest of all the sequoia groves; the effects of the 1878 Timber and Stone Act, however, would ultimately force him to change his mind. One of the provisions of the act was that land could not be sold until it was surveyed by the federal government and then broken into mile-square units under the township-and-range system.⁶ Surveyors working under contract for the General Land Office laid the requisite lines across the Tule River headwaters in the early 1880s, and by 1884 the organization's Visalia District office reported that a timber rush was under way, with twenty-three private individuals having filed claims in the area. Most of these filings included giant sequoia land, and by the time the rush was over essentially all of the Dillonwood and Mountain Home sequoia groves had gone into private hands.⁷ Half a dozen mills soon opened, and until giant sequoia logging wound down in the Tule River groves after 1905, these mills would produce at least ninety million board feet of lumber, much of it from the region's sequoias.⁸

(To understand the scale of what this chapter explores, we must divert briefly from our historical narrative and explain a measurement used in the timber industry. The primary measurement we will use in what follows is a "board foot" of lumber, which, simply put, is one linear foot of board measuring one foot wide and one inch thick—what a carpenter would call a 1x12. A board ten feet long with these dimensions would thus equal ten board feet. But how do we grasp the concept of *millions* of board feet? One way is to imagine laying the boards end to end and seeing how far they would extend. Using this logic, a million board feet laid upon the ground would stretch 189 miles!)

How many sequoias fell in the Tule River country in these years is hard to say, but it is clear that the cumulative impact of the lumber industry changed these forests profoundly. More than a century later, countless huge stumps still dot the landscape. The damage was so extensive that in 1900, when John Muir reworked his earlier writing in anticipation of publishing *Our National Parks*, he dropped the description of the Tule River forests as being the finest of all the sequoia groves, modifying his text in a way that shifted this praise to the still-uncut Giant Forest grove in the Kaweah watershed.⁹

Meanwhile, on the ridges along the south rim of the great canyon of the Kings River, things were even worse for the Big Trees. The relative accessibility of these groves made them perfect targets for acquisition under the Timber and Stone Act. Since the closure of Hyde's Mill at the end of the 1877 season, the region had seen a respite in giant sequoia logging, but that ended six years later when Smith Comstock erected his steam-powered mill in the heart of the Big Stump Basin grove. The forest took its name from a massive long-dead sequoia snag known then as "Old Adam," but over the next several years the landscape blossomed with hundreds of new stumps. Comstock's operations at Big Stump represented at the time the most intensive sequoia logging yet seen in the Sierra. Large as it was, Comstock's operation still reflected traditional approaches to logging, and it would be one of the last large operations to be carried out with oxen dragging the massive sequoia rounds to the mill along wooden skidways laid in the ground. Workmen greased the skids as the logs squealed their way the relatively short distance to the mill.[10] Comstock's axe men cut all but the largest of the Big Trees on the site, and when he wrapped up in late 1888 only a few very large sequoias remained—trees that had been too large to fell efficiently because of the biggest trees' tendency to shatter badly when they hit the ground. (Others would eventually come back and harvest the few trees Comstock had left behind.) Later studies of the site concluded that at least 350 mature sequoias, and perhaps as many as 500, had been felled at Big Stump.[11]

A few miles away, even bigger schemes were already under way. In the summers of 1886 and 1887, Timber and Stone Act claims in the sequoia groves along the south rim of the Kings River Canyon spread quickly. The nature of what was planned took a little longer to come out, but in March 1888 the *San Francisco Chronicle* published a story revealing that the men behind the large-scale land claims were Hiram Smith and Austin Moore, figures whose names were also popping up as purchasers of land in the Tule River groves. Smith and Moore were no strangers to the lumber business, being the owners of 18,000 acres of timberlands in Washington State and a large lumber business in Stockton, California.

But what they had in mind this time was ambitious even for them. Historians would estimate later that Smith and Moore used the Timber and Stone Act to take control of more than forty square miles of forested acreage in and around the Kings River sequoia groves.[12] Twenty-first-century minds boggle. How could the government essentially give away such huge tracts of forest for almost nothing? The answer is sadly simple: such was the prevailing ethic of late-nineteenth-century America, which believed the purpose of the public lands of the West was to enrich the nation by enriching individuals.

Knowing that they could thus count on access to a large volume of timber, Smith and Moore organized the Kings River Lumber Company, which had a capital value at its inception of $1 million, an enormous sum at the time.[13] Soon, the new company was pouring money into the woods. By the following summer, the enterprise had two sawmills, and a newly erected dam was in the process of turning Mill Flat (the previous site of Thomas's Mill) into a reservoir. A fifty-four-mile-long flume opened in September 1890 to connect the mills with the Southern Pacific Railroad at Sanger, in the San Joaquin Valley. In the four months after the flume opened, about twelve million board feet of cut lumber sailed down the long wooden trough to the lowlands.[14]

The woods immediately around the two new mills offered mostly pine and fir trees, but in 1891, Smith and Moore began constructing a narrow-gauge railroad toward the sequoia groves they had acquired, an endeavor that required, among other things, the importation to the site of a locomotive weighing thirty-six tons. It came up the mountain in pieces carried by wagon. By July 1893, the "Sequoia Railroad" had been extended five miles to the edge of the Converse Basin, the largest single sequoia grove in the Sierra. At this point, however, the economic panic of 1893 had hit, and by the time the dust settled in August 1895, Smith and Moore had lost effective control of what had now been renamed the Sanger Lumber Company, after the lowland community where the flume ended.[15]

Too much money had been spent to stop now, however, so the new owners poured yet more cash into the concern, with the result that the railroad was extended into the heart of Converse Basin and another sawmill constructed there.[16] The new mill began cutting sequoia logs on June 30, 1897.[17] The largest sequoia logging operation ever to be conducted had begun. Before the mill finally closed a decade later, the Converse Basin Grove was gutted of old-growth sequoias. No one knows exactly how many sequoias fell in Converse Basin, but we do know that the mill produced more than two hundred million board feet of saw timber between 1897 and 1907, with historians estimating that one-third to one-half of that volume came from the Big Trees.[18]

The Converse Basin sawmill of the Sanger Lumber Company facilitated the destruction of one of the Sierra's largest giant sequoia groves between 1897 and 1907. COURTESY OF NPS, SEQUOIA AND KINGS CANYON MUSEUM COLLECTION

The logging of Converse Basin broke ground in other ways. Befitting the enormous investment made in the operation, the Kings River and Sanger Lumber Companies became show places for new technology. Instead of oxen hauling the felled trees, steam winches known as "donkeys" powered continuous loops of woven wire rope that pulled the huge logs along greased skidways, some of them several miles long. In the mill, enormous looped bandsaws sliced the largest logs into manageable pieces under the glow of electric arc lights. In the woods, axes were still employed to undercut the trunks, but long saws, some of them up to twenty feet in length, actually felled the trees and then cut the fallen logs into short sections so that the steam winches could pull them to the mill.*

Photographers came to the grove year after year to record the work, and the resulting images circulated widely. Loggers posed before sequoias with specially crafted undercuts that emphasized both the scale of the

---

* To fell a standing tree it is necessary to make both an undercut and a backcut. The undercut determines the direction the tree will fall, and the backcut severs the remainder of the trunk and allows the tree to fall into the undercut. In photos that show the felling of giant sequoias, the enormous cuts being made by hand are the undercuts.

trees being felled and the obvious skill and strength of the men who felled them. Even today, the photos carry an emotional context of heroism—of American ingenuity and know-how.[19] At the same time, of course, they also speak of environmental tragedy.

In the end, only one thing turned out to be missing in Converse Basin: profit. The huge initial investment in the form of steam sawmills, the railroad, and the fragile flume that transported the cut lumber fifty-four miles from mill to market crippled the company throughout its existence. Because of the brittle nature of the big sequoia logs, vast amounts of wood were lost to breakage and wastage. The original Kings River Lumber Company collapsed into bankruptcy in 1894, and its successor, the Sanger Lumber Company, did little better, losing money year after year.

Several decades later, Walter Fry and John R. White would somberly sum it all up in their book *Big Trees*:

> There must be millions and millions of board feet of lumber in boles and limbs of sequoias lying on the ground in the Converse Basin; and there are millions more in the chutes and trestles upon which the logs were conveyed to the mills....There are miles and miles of such trestles and chutes—miles and miles and miles. Waste, waste, waste....Probably not one-half of the trees destroyed, of all species, ever reached the mill to get converted into lumber; quite possibly not one-third.[20]

Yet the destruction would continue well into the twentieth century. With so much money invested, the operators had no choice but to try to make it work. In 1907, fresh investors came in, hoping to succeed where their predecessors had failed. They relocated the mill, built another reservoir (modern-day Hume Lake), another mill, and another railroad, and extended the flume. Then they resumed cutting sequoias, still losing money season after season.[21] Logging continued sporadically in the groves into the 1920s before continued financial losses finally exhausted the last of a long series of unfortunate investors.

Sadly, the financial experience of the Sanger Lumber Company and its corporate successors proved the norm rather than the exception. Despite easy access to almost-free land, no one ever struck it rich cutting down and milling giant sequoia trees. The combination of factors that defeated the

For those involved, the logging of the Big Trees at the beginning of the twentieth century was a heroic enterprise. Later generations have not been so sure. COURTESY OF NPS, SEQUOIA AND KINGS CANYON MUSEUM COLLECTION

lumbermen in Converse Basin—remote location, high expenses, enormous material losses to breakage, and a mediocre final product—simply could not be overcome. Ironically, small mills did the best over time, economically speaking. An individual could sometimes make a modest living taking down a sequoia now and then and breaking it apart into fence posts and grape stakes. Big operations, on the other hand, always lost big money. It was as if the trees, in some anticipation of the arrival of industrial logging, had figured out how to defeat it before the lumbermen ever arrived.

Fortunately for posterity, not every story that starts with the Timber and Stone Act turned out as disastrously for the trees as did the one that took place in Converse Basin. For a change of pace, let us now turn to the sequoia groves in the heart of the Kaweah River watershed, including the Giant Forest. On October 5, 1885, a group of thirty-seven men walked into the General Land Office in Visalia and filed claims under the provisions of the Timber and Stone Act. All were members of the San Francisco–based Co-operative Land Purchase and Colonization Association, a utopian society seeking to escape the mainstream economy and take control of their own destinies. At the end of the month, additional members of the group filed sixteen more claims.[22] All together, fifty-three individuals from the association filed claims on more than 8,000 acres of Sierra forest within the Kaweah watershed. At the center of this land lay the Giant Forest—the sequoia grove that had so excited Muir during his 1875 reconnaissance.

The fact that idealistic utopians could see a future based on intensive logging demonstrates how widely accepted large-scale logging was by nineteenth-century Americans. One did not have to be a robber baron to cut down forests, and indeed profits from lumber could be imagined as many things, including freedom. Allowing individuals to pursue such a dream had been a part of the original concept that became the Timber and Stone Act, even if the resulting reality granted its benefits primarily to large corporations.

The endless tension between corporate America and those who sought to take another path may explain what happened next. Throughout the Pacific Coast states, grants of land under the Timber and Stone Act, many of them obviously fraudulent, usually proceeded unchallenged. But in the case of the utopian community, the staff of the Visalia Land Office balked in the face of what they decided must be yet another attempt to abuse the statute. Instead of processing the claims, they reported their concerns to officials in Washington, DC, and by the time the members of the association returned after the statutory sixty-day waiting period to pay their title fees and take possession of their claims, they found a surprise waiting for them. By order of General Land Office commissioner William Sparks, the tracts in question had been "withdrawn from entry" (that is, declared no longer available under the provisions of the Timber and Stone Act) pending an investigation to see whether the sudden surge of filings represented fraud.[23]

The officials in Visalia had their reasons. They noted that all the filers in this group listed the same address in San Francisco on their individual paperwork, and they speculated that these working-class men were likely not the actual claimants since they obviously did not have access to the funds necessary to build a road or flume or railroad to the grove, something that would have to be done if the area were to be logged. An investigator from the General Land Office arrived in Visalia in December and reinforced these concerns.[24] The claims remained in suspension.

Confident nonetheless that their claims would ultimately be granted, the members of the association, who had by now reorganized themselves as the Kaweah Colony, moved forward. They knew that they did not represent a greedy corporation intent upon subverting the law. Indeed, they represented the very antithesis—a group of individual citizens banding together for their mutual benefit. Undaunted, they explored the possibility of building a railroad to the Giant Forest, an undertaking that upon study proved to be prohibitively expensive owing to the rugged terrain. By the fall of 1886 they had begun instead the construction of a wagon road up to the Big Trees, an effort they would continue for the next four years.[25] By the summer of 1890, the colonists' road-building effort had brought them to the edge of the merchantable timber and within a few miles of the Giant Forest itself. They began logging that summer, even though their Timber and Stone Act claims still remained in limbo. Had they been a corporation, we can suspect they would have been allowed to proceed unmolested. But they were not, and the events that followed were about to sweep over them with disastrous results. But that's a story for a later chapter.

The great logging boom that engulfed the sequoia groves of the Sierra Nevada following the passage of the Timber and Stone Act of 1878 provides powerful insights into how perceptions of the sequoias had evolved since their effective rediscovery in 1852/53 by Euro-Americans. Initially the trees had been wonders, almost unimaginable in their size and assumed age. In the beginning, even those who intended to use the trees to make their own fortunes saw them as objects to be exhibited as horticultural marvels, a species whose seed justified rushing halfway around the world to share the news. Literature and tourism had followed, with the trees quickly gaining worldwide fame. Within a dozen years, the

Mariposa Grove—one of the two best-known of the groves uncovered in the 1850s—had been set aside by the federal government in what would become the first approximation of our modern national parks.

But as the southern Sierra was explored and it became apparent that there were many more trees than originally perceived, other perspectives arose. In the 1880s, the giant sequoias of the southern Sierra Nevada became a commodity—materials that could be harvested and sold for personal gain. And the passage of the Timber and Stone Act made it all possible. The law included no provision for setting aside resources of special value, and indeed the assumptions that supported the legislation clearly believed that privatization and harvest represented the best national use of western forests. Some Californians held other opinions—this was, after all, the state that had passed a law prohibiting the cutting of sequoia trees larger than ten feet in diameter—yet, during the sequoia-logging boom that extended for the next two decades, no one ever attempted to apply this restriction. And it was not just that attempts to protect the trees via this avenue failed; there was simply no discussion of the law whatsoever. It had become a dead letter, ignored and forgotten.

If lumbermen believed that they could make money cutting down giant sequoias, then the trees must fall. Federal legislation had made them, in essence, free for the taking.

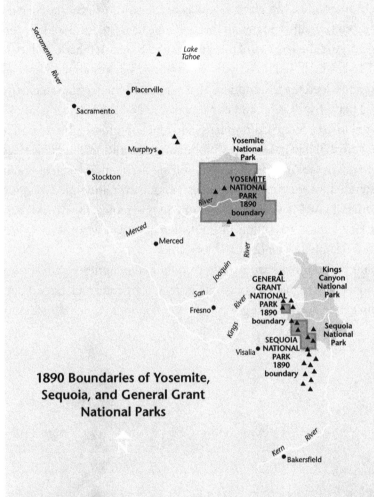

CHAPTER EIGHT

Of Tunnel Trees and National Parks

On a cold November afternoon, I stand before what may well be the most famous tree in the world. The problem, unfortunately, is that despite its continuing fame, it broke apart, collapsed, and died nearly half a century ago. Its death, however, has done surprisingly little to dampen the tree's renown. For people all over the world, decades-old images of the Wawona Tunnel Tree still define the giant sequoias.

I have one of those images, a black-and-white photo taken in the years just before the First World War. My father's mother, then a young nurse touring the West between jobs, sits atop a horse-drawn stage that has just emerged from a tunnel cut through a massive sequoia trunk. All the visitors captured in the photo, my grandmother among them, pose with that Edwardian formality that now seems so distant. Even the horses seem to sense the need for dignity that drove the era. But the tree brings a gravitas to the scene that makes all the humans at its base seem small and insignificant. The sequoia's massive and seemingly immutable trunk reduces the stagecoach to a mere plaything. That, of course, is the enduring genius of this manmade setting.

No one knows how many times the Wawona Tunnel Tree was photographed during the nearly nine decades that it stood as the best-known feature in the Mariposa Grove, but the total must run into the tens of millions. That the fame of the Tunnel Tree eclipsed even that of the nearby Grizzly Giant, a much larger tree, confirms just how successful a marketing idea this attraction became.

A similar message can be drawn from the fact that visitors to all the sequoia groves of the Sierra still search for it, still assume that something so iconic must endure. In Sequoia National Park, where I worked for decades, the question became a staff joke. "Where is the tree you can drive through?" the visitors asked, day after day, summer after summer. That it had fallen in 1969 made no impact on them at all. "No," they would respond, "not the tree in Yosemite. I am looking for the tunnel tree *here*, the one my parents drove me through when I was just a child." We heard the question so many times that after hours we worked out answers we could never use in public. "Where is the tree you can drive through? Well, sir, that depends on how fast you are driving!" Yosemite's rangers still wrestle with the problem today, a half century after the tree's demise. On this November afternoon, I am amused to note that the tree has been subtly renamed. All the park's maps, trail signs, and exhibits now use a modified title; the tree has become the "Fallen Wawona Tunnel Tree." I wonder if it makes any difference.

Decades after its death, the moldering remains of the Wawona Tunnel Tree remind us of a critical moment in the history of the Big Trees. By the 1880s, with the natural extent of the trees' range fully defined, and with large-scale logging taking off in multiple locations, Muir's public worries about the future of the groves were coming to a head.

As sequoia logging took off, it was apparent that if the Big Trees were to have any future as tourist draws they would need to be better marketed. Even by the late 1870s, the giant sequoias of the Sierra Nevada had lost much of their novelty. The initial excitement of their discovery had passed, and they became expected rather than extraordinary. Lithographs and photographs of the trees had circulated widely, and they had been the subject of numerous newspaper and magazine articles. Giant samples of their wood and hollowed-out re-creations of their bulk had been put on display on several continents. On the ground, of course, the sequoias remained immensely impressive, but increasingly they had to

The tunnel tree phenomenon began in 1878 with the piercing of the Dead Giant along the Big Oak Flat Road to Yosemite Valley. The idea would capture the public's mind and be widely copied. COURTESY OF YOSEMITE NPS LIBRARY

compete with other California attractions. How to draw visitors to the groves became a major question.

The drive-through idea arose not at Mariposa or Calaveras but instead in the small Tuolumne Grove a few miles northwest of Yosemite Valley. The grove had no sequoias of exceptional size and had remained mostly unknown to the world into the 1870s.[1] The area gained new prominence, however, when in 1874 the Big Oak Flat Road, one of the two wagon roads built into Yosemite Valley from the northwest that year, passed through the grove. The competing Coulterville Road also ran through a stand of sequoias—the nearby Merced Grove—and in 1878, seeking to make their Big Tree experience more memorable, the operators of the Big Oak Flat hit upon what would turn out to be a brilliant marketing idea. That summer, workmen using augers began widening a natural fire scar in the Dead Giant, a sequoia snag located barely one hundred yards from the grade of their toll road. By the time they were done, the workmen had opened a path through the snag large enough to allow the passage of a tourist stage.

The tunnel proved an immediate hit. A photographer captured the first stagecoach to pass through the tree, and the image was widely reproduced.[2] The insertion of a horse-drawn stage, an object everyone knew well in the nineteenth century, into the basal portion of a giant sequoia tree brought the tree's size into focus for many who had remained skeptical

The Wawona Tunnel Tree in the Mariposa Grove would capture public attention like no other. Here, Galen Clark stands by the iconic opening. COURTESY OF YOSEMITE NPS LIBRARY

in the face of other Big Tree images. The tunnel image "clicked" in a way no previous photo or lithograph had.

Meanwhile, hotelier Henry Washburn, whose family now controlled the southwestern approaches to Yosemite, continued to fight for a share of the tourist flow headed to Yosemite Valley. The Washburns had finished their road to the famous valley in July 1875, and they also commanded tourism in the nearby Mariposa Grove, by far the best of the region's three sequoia groves. After 1879, with their new hotel open at Wawona and a wagon road extended into the grove, the Washburns felt that they were offering their visitors a strong sequoia experience. Yet something was still missing: they did not have a tunnel tree. They corrected this shortfall in 1881, when they selected a living tree in the uppermost part of the grove and had a tunnel cut through it for the loop road they had laid out within the grove. The result easily surpassed the tunnel through the Dead Giant along the Big Oak Flat Road. This new tunnel tree was no fire-charred snag but instead a vigorous living sequoia.[3] And it was surrounded by the glories of the upper portion of the Mariposa Grove.

By the late 1880s, tunneled trees had become a must-have for many tourists to the giant sequoias. The owners of the Calaveras Big Trees responded by cutting a passage through the Pioneer Cabin Tree, and in 1895 the Washburns went so far as to tunnel through a second tree in the Mariposa Grove—one that could be accessed more easily during the winter.[4] The image of a wheeled vehicle passing through a giant tree had by now been firmly established in the public mind as the defining popular vision of the trees. It was a vision that would have a long life.

None of this aggressive tourist marketing of the Big Trees, however, offered much in the way of enhanced protection for the many sequoias scattered across the public domain. Instead, as the 1880s progressed, the provisions of the Timber and Stone Act continued to open grove after grove to logging. Most Californians had no real problem with these developments, but small pockets of concern began to form. Two of these evolved into efforts to protect actual trees, and within a few years they would lead to the creation of three new national parks—the second, third, and fourth units of the American national park system. The two campaigns are often confused or conflated, but they began independently and had separate initial goals.[5] The fact that both endeavors came to fruition around the same time—in the late summer and early fall of 1890—was the result of coincidence rather than cooperation between the participants.

The northern group, made up of individuals living mostly in the San Francisco Bay Area, focused on the Yosemite region. A more southerly group of Big Tree advocates came together in Tulare County, literally almost in the shadow of the mountains that held the biggest groves and largest individual trees. The southern group had begun their efforts a decade earlier when Frank J. Walker and Tipton Lindsey of Visalia began lobbying for the withdrawal from sale of the Grant Grove, an area used by Visalians as an escape from the summer heat of the San Joaquin Valley.*
This was successful, and in January 1880, Theodore Wagner, the federal surveyor general for California, wrote to the General Land Office agent in Visalia and asked that he withdraw from sale a four-square-mile section of land in and around the grove. The intent was that Congress would follow

---

* This is the same grove previously referred to as the Kings River Grove.

up with more formal arrangements to protect the land.⁶ Although that did not happen immediately, for the remainder of the decade those four square miles remained unavailable for sale, and the trees in this small area remained under government ownership while the neighboring Big Stump and Converse Groves were sold into private hands and began to be logged.⁷

Meanwhile, in the San Francisco Bay Area, another group began to coalesce. Almost from the beginning, it had been apparent to those who cared about the beauties of the Sierra that the land grants made in 1864 to protect Yosemite Valley and the Mariposa Grove needed to be expanded. In March 1881 the state-appointed commissioners of the Yosemite Grant passed a resolution calling for an enlargement of the reserved area, but ultimately nothing came of it. Other efforts to increase the amount of land protected under the Yosemite Grant sprung up in 1884 and 1886, but again the bills did not survive.

Not until the end of the decade did the conservation movement gain fresh ideas and energy, and by coincidence, forward movement occurred on both fronts. In the south, Visalia newspaper editor George Stewart began editorializing in April 1889 on the state of the groves. His immediate focus was the permanent protection of the four-square-mile area surrounding the Grant Grove. Attempts were under way to end the suspension of sale so the land could be transferred to private ownership. Stewart and several friends wrote to the national commissioner of the General Land Office requesting that the lands remain withdrawn from sale, and they were told in response that if the withdrawal were to be made permanent, Congress would have to take action.

This response put them to work. They set out to convince Congress to protect not only the Grant Grove but also those other sequoias in Tulare County that had not yet been claimed under the Timber and Stone Act.⁸ When they looked into this issue, they discovered that the largest area of giant sequoia forest remaining in Tulare County that had not been compromised by Timber and Stone Act claims was a steep mountainside complex of trees along the South Fork of the Kaweah River, known today as the Garfield Grove. Accepting this information, they began working with William Vandever, a congressman from central California, to move a bill to set aside the Garfield Grove as a national park. To increase support for their proposal, they argued that preserving the forests of the Sierra Nevada from logging would enhance Central Valley agriculture by protecting the watersheds that produced irrigation water for summer crops. It also helped that, when compared to the region as a whole, the area in question was relatively small, consisting of only two townships of thirty-six square miles each. Since there were no timber claims in the way, the bill made good progress. After only minimal consideration and with no one from California objecting, Congress passed the bill for a federal "forest reserve" to protect the sequoias and then forwarded it to President Benjamin Harrison, who signed it on August 25, 1890.[9]

At almost exactly the same time, the northern effort to protect the Big Trees gained traction as well. The on-again, off-again attempt to enlarge the area of the Yosemite Grant gained new energy in the spring and summer of 1890, but with a twist. John Muir and others had been complaining about state management of Yosemite Valley and the Big Trees for years, documenting numerous small and not-so-small abuses. Muir had begun working with *Century Magazine* publisher Robert Underwood Johnson in June 1889 in an effort to make fundamental changes at Yosemite, and these efforts eventually paid off. In March 1890, Congressman Vandever also introduced a bill to expand the protected area in the Yosemite region, but it called not for transferring more land to the state reservation but rather for the creation of a new national park to surround the existing state-controlled reservations.[10] Muir and his northern friends were delighted, even if the bill did not cover as much land as they wanted to see protected.

Muir wrote two powerful essays for the *Century* that summer, and the chances looked good for Vandever's Yosemite bill. Then, on September 30, with essentially no public notice, Vandever withdrew his bill and allowed a substitute to be entered in its place. This new bill moved quickly through Congress and was passed and sent on to President Harrison, who signed it on October 1. Although it still focused on the Yosemite region, the replacement bill took quite a different direction than Vandever's earlier draft; these changes would revolutionize the status of the giant sequoias.

The full origins of the second California national park bill of 1890 remain a mystery. Even today, we do not know for certain who drafted it, exactly how it came to be substituted for Vandever's earlier Yosemite bill, or what was behind the change. What we do know is that the substitute bill took bold steps with regard to both national parks and giant sequoia trees. The law of October 1, 1890—formally *An Act to Set Apart Certain Tracts of Land in the State of California as Forest Reservations*—contained two critical passages. First, it created a large tract of "reserved forest lands" that was to be managed by the secretary of the interior so as to preserve its natural features for public enjoyment. This new reservation, which would surround the land protected by the Yosemite Grant, measured more than 1,500 square miles; in contrast, Vandever's original Yosemite bill had proposed a reservation barely 20 percent as large.[11]

The second critical section of the bill addressed the sequoias in Tulare County. A week earlier, two townships of mountain land had been set aside to protect sequoias within the Kaweah River watershed, and to that area this new bill added almost five more townships for a total of over 170 square miles. Contained within these lands were two dozen of the sequoia groves Muir had explored during his 1875 wanderings. The bill also gave permanent "forest reserve" status to the four-square-mile tract that had been temporarily withdrawn from potential sale in 1880 at the Grant Grove.

Historians have long speculated about how this legislation came to be. None of the principals at the time, including Muir and Stewart, admitted to even knowing about the substitute bill until well after the legislation was signed into law. But there are clues. The strongest of these is a map that came to light in the National Archives in the early 1960s. Drafted in San Francisco by the Southern Pacific Railroad on October 10, 1890, the map clearly demonstrates that the railroad knew what was going on well before the rest of California learned what Congress had done.[12] Was the railroad, then the most powerful corporation in California, behind the substitute bill? The supposition has long been that it was.

There is strong evidence to suggest Southern Pacific involvement in the second California national park bill of 1890. As both a promoter of California tourism and a major player in the California timber market, the Southern Pacific had good reasons to support the withdrawal of these lands from potential logging and their diversion toward tourism. During the critical period in September 1890, Daniel K. Zumwalt, the railroad's district land agent for the San Joaquin Valley, was not only present in Washington, DC, but actually being hosted there by Congressman Vandever. Moreover, Zumwalt knew the Sierra well and had begun his stay in Washington by lobbying for Vandever's original Yosemite bill.[13]

One more curious detail suggests the influence of the railroad. The acts of September 25 and October 1, 1890, did not specifically create national parks but, instead, used the terms "public park, or pleasure ground" (September 25) and "forest reserves" (October 1). Not until October 21, when the secretary of the interior, John Noble, formally issued regulations for the management of the new reservations, were they given proper titles, and rule number one was that they would be known as "national parks." The names chosen by Noble were Yosemite National Park, General Grant National Park, and Sequoia National Park.[14] What's curious is that somehow the Southern Pacific knew all of this before it happened! The railroad's map of October 10, drafted ten days before the parks were officially named, identified the new reservations in the Kaweah watershed as Sequoia and General Grant *National Parks*.[15]

Certainly, tourism played a role in the railroad's thinking. In 1886, the Southern Pacific had built a branch line to a new Sierra foothill terminal it named Raymond. From there, horse-drawn stages took visitors to the Mariposa Grove and on to Yosemite Valley. From that time into the middle years of the twentieth century, the railroad would regularly promote visiting Yosemite Valley and the Big Trees.[16]

In reality, the creation of Yosemite National Park had only a small beneficial effect upon the status of the Big Trees. The new national park surrounded the state reservation that had provided a form of protection to both Yosemite Valley and the Mariposa Grove since 1864, and within the boundary of the original grant, things went on as before. The new park did provide legal protection for both the Tuolumne and Merced Groves, but

---

* In calling these new reservations national parks, Secretary Noble obviously chose to model them on Yellowstone, which was at that time the nation's only previous national park. On the same day that Noble christened the new parks, he also wrote to the secretary of war requesting that the army detail troops to the parks to protect them, again building on the Yellowstone model, where troops had been present since 1886.

these relatively small tracts of sequoia forest had already been practically exempted from logging by their de facto designation as tourist features along the two northern wagon roads leading into Yosemite Valley.

The creation of the two southern parks had more significance for the sequoias, however. The designation of General Grant National Park brought permanent protection to the General Grant Tree and the other monarchs that immediately surrounded it. These trees would now survive even as the nearby and much larger Converse Basin Grove fell to the axe. It was the creation of Sequoia National Park, however, that most strongly advanced the cause of tree protection. Although only about one-sixth the size of the newly created Yosemite National Park, Sequoia was far richer in sequoia trees. While Yosemite focused primarily on the High Sierra, Sequoia had been laid out to maximize the protection of its Big Trees. Within its boundaries fell twenty groves of sequoias, including the incomparable Giant Forest that had so excited Muir fifteen years earlier. The future of the Giant Forest remained problematic, however, for the Kaweah colonists had blanketed the grove with Timber and Stone Act claims back in 1886, and although those claims had been suspended and the colonists had yet to cut any trees in the grove, the claims were still on the books. Many of those who had worked to create the Tulare County park, including George Stewart, assumed that title to the Giant Forest would rightfully be conferred to the utopians. After all, in filing their claims they had committed no fraud, a fact confirmed by several investigations.

Affairs quickly turned against the colonists, however. A new investigation of the situation by yet another General Land Office agent concluded without much evidence that "presumptive fraud" was indeed in effect, and on November 30, United States marshals arrested four of the colony's trustees, charging them with the crime of "timber trespass"—that is, of cutting trees they did not own. The following April, Secretary Noble cemented the case for the prosecution by verifying that since title to the lands in question had not yet been transferred, and since Congress had made no provision for the colonists, their claims were without merit. The Giant Forest would become a part of Sequoia National Park.[17]

Would the Timber and Stone Act claims of the Kaweah Colony have been rejected so casually had they been filed by a powerful corporation?

We can only speculate, but history suggests that the administrative system of that time felt much more comfortable supporting big lumber companies than social experimenters. Certainly, the legal apparatus of the time felt no sympathy for these utopian lumbermen.

Cavalry troops arrived at Sequoia, General Grant, and Yosemite the following May, and the three new national parks settled down to a quiet existence. Congress appropriated no funds for the development of the new reservations, so activity by the federal government was limited to fighting fires, preventing logging, suppressing hunting, and expelling trespassing herds of domestic sheep. The presence of a developed tourist industry focused on Yosemite Valley ensured that the surrounding Yosemite National Park received a flow of visitors. The two southern parks remained relatively unknown. A wagon road led to General Grant National Park, and it became an army-managed summer campground for local residents, and in Sequoia National Park, only the road to the mines at Mineral King passed through any part of the park, and the rest of the reservation remained essentially undeveloped and infrequently visited.* What counted most, however, was that the lands within this new reservation were no longer for sale.

The last years of the nineteenth century saw the position of the giant sequoias changing in American society. Big Tree marketing and tourism—as exemplified by the tunnel trees—had reached new highs, and the role of the groves as special magnets for travelers continued to grow. The creation of the three giant sequoia national parks in the autumn of 1890 confirmed this continuing shift of attitudes—including the perception that the Big Trees were fellow living things. Yet even as federal troops settled in to police General Grant National Park, mammoth trees continued to fall to loggers only a few miles away in Converse Basin. Even more ironically, the quest to find and exploit exhibit trees continued as well.

In the spring of 1891, at the same time federal cavalry troops arrived in the new national parks, a team of workmen collected at the base of the

---

* The army did not maintain the road constructed by the Kaweah Colony, and it rapidly fell into disrepair.

Mark Twain Tree in Big Stump Basin. Lumberman Smith Comstock had determined the tree to be too large to cut and had left it standing when he had wrapped up his logging three years earlier, but now the very magnitude of the tree's bole was seen as an asset rather than a problem.

Money for the project came from none other than Collis P. Huntington, one of the owners of the Southern Pacific Railroad. By the early 1890s, Huntington had amassed an enormous fortune, and in addition to pursing new business investments, he had begun to collect art and support museums. It was in this latter role that Huntington sent men to fell the Mark Twain Tree. As a key figure in the management of the Southern Pacific, Huntington likely knew of the corporation's lobbying the previous fall for the creation of the three Sierra Nevada national parks. Now, barely six months later, he was funding a project to fell one the largest remaining unprotected sequoias so that its cross sections could be provided to both the American Museum of Natural History in New York City and the British Museum.[18]

A century later, it is impossible to discern whether Huntington noticed the irony of his plan. Within a half mile of the boundary of a new national park his corporation had just helped create, Huntington happily paid for the destruction and removal of a sequoia twenty-five feet in diameter and more than 1,300 years old. His railroad even provided free shipping of the cross section from California to New York City.[19]

Based on the historical record, it is clear that during the summer of 1891 the felling of the Mark Twain Tree generated much more excitement than the beginnings of on-the-ground protection for neighboring General Grant National Park. Pioneer photographer C. C. Curtis documented the destruction of the Big Tree in excruciating detail. Even today, his photo of the huge tree crashing groundward is one of the most powerful and often reproduced giant sequoia photos of the pioneer era. Before the year was out, the photo of the huge falling tree would be converted to an engraving and published for the enjoyment of an admiring national audience in *Scribner's Magazine*.[20]

And the irony did not stop there. The following summer, yet another large sequoia came down for exhibition purposes, this felling sponsored

C.C. Curtis's photos capture the process that converted the General Noble Tree into an exhibit at the 1893 World's Columbian Exposition in Chicago. First, the upper portion of the trunk was cut away. Then the lower portion of the trunk was hollowed and sectioned.
COURTESY OF NPS, SEQUOIA AND KINGS CANYON MUSEUM COLLECTION

by no less than the United States Department of the Interior, the federal organization commissioned to oversee the Sierra Nevada's national parks. The World's Columbian Exposition of 1893, a celebration of the four-hundredth anniversary of the arrival of Columbus in the New World, provided the impetus.[21] Interior Department agents scouted the sequoia groves around General Grant National Park and eventually settled upon a massive monarch tree a few miles to the north of the park, on the edge of the Converse Basin Grove. Honoring their ultimate boss, they named the soon-to-be-felled tree for General John W. Noble, the secretary of the interior.

The tree grew on land that had already been sold by the government to Smith and Moore of the Sanger Lumber Company and, in fact, even

As the General Noble Tree was hollowed out, the resulting sections were loaded onto wagons to begin their long journey to Chicago. COURTESY OF NPS, SEQUOIA AND KINGS CANYON MUSEUM COLLECTION

as preparations were made to destroy this monarch, the lumber company was constructing a logging railroad that would enter the Converse Basin within sight of the General Noble.

Again, photographer C. C. Curtis documented the process. Because the government wanted to extract for exhibit a hollow tube of giant sequoia trunk two stories high, the loggers first erected a scaffolding fifty feet above the ground. Working at that height, they felled the tree, allowing Curtis another opportunity to capture a sequoia in the act of crashing to the ground. With the top of the tree now out of the way, the workmen hollowed and sectioned the tall standing stump, reducing it to pieces that could be loaded into wagons and carried away. By the end of August, only a stump remained.[22]

The following summer, in Chicago, the hollowed trunk of the General Noble Tree stood in a place of honor in the center of the rotunda of

After its short time at the World's Columbian Exposition, the hollowed-out trunk of the General Noble Tree was set up on the National Mall in front of the Smithsonian Institution in Washington, DC. It remained there, an incongruous object, for decades until it was removed in the early 1940s.
COURTESY OF NPS, SEQUOIA AND KINGS CANYON MUSEUM COLLECTION

the fair's Government Building. Modern interpretations of this placement provide context for readers to consider:

> Prominently situated in one of the most notable fair buildings, the trunk asserted the nation's arrival as a mighty power, one whose promise and latent riches were as massive as the tree's height and bulk.[23]

As displayed at the World's Columbian Exposition, the sequoia reflected not conservationist concerns about the fragility of the earth's environmental features but rather the confidence of a nation generously endowed by nature and emerging as a world power. If God gave man these wondrous trees, man had no qualms doing with them whatever he wished.

CHAPTER NINE

For the Greater Good

The trail climbs steeply, and it doesn't take long to leave behind the noise of the highway and the drifting voices of visitors marveling at the Big Trees that grow in the heart of the Giant Forest of Sequoia National Park. Despite the fact that the trail has been here for more than a century, few ascend this particular route. The uphill lasts only a few hundred yards, but the trail climbs aggressively enough that soon I am looking out into the crowns of trees whose bases I passed only a few minutes ago. I stop and catch my breath, enjoying as I do the superlative stands of sequoias that grow here.

Like many trails within the Giant Forest, this one possesses a historic name. Maps still label it as the "Soldiers Trail," although few who walk the trail today know who the soldiers were or why they were here. My local trail map offers only one hint: on the other side of the hill I am climbing, a trailside flat is identified as "Soldiers Camp," and when I first visited this site in the 1970s, lingering evidence of its past habitation could still be found. Dry boards lay scattered among the trees, and here and there one could still find telephone insulators attached to trees. Now, several decades later, even those faint traces have faded away. Almost nothing remains to separate this site from the surrounding forest.

The names Soldiers Camp and Soldiers Trail endure to tell us that for a few short years this was the residence of the troops assigned to protect and administer Sequoia National Park. I find a fallen log and make myself comfortable, taking in the quiet and letting my imagination try to conjure up the energy that must have infused this site for the three summers it was occupied by the military.

Cavalry troopers from Company A of the First Cavalry opened this camp on June 23, 1911, and in the two succeeding summers other companies from the First Cavalry occupied the same locale, which they called "Camp Sequoia." The army made only a small investment here, housing the soldiers in wood-framed tents; given the Sierra's benign summer climate, much of the business of living and working here must have taken place out of doors. To supply the camp, the army did build a water line some 6,000 feet in length, as well as a trail to connect the site with the popular tourist camping areas a mile to the north. That trail, with its steep grades designed by horsemen, brought me here today. And it tells a story, if we take the time to listen.

The creation of the three Sierra Nevada national parks in the fall of 1890, an action taken quickly and with no significant consideration of what it might entail to sustain the parks, opened a new chapter in the saga of the sequoias. Up to that time, little attention had been given to the question of what might be involved in managing giant sequoia trees for long-term preservation. Most people assumed that since the trees had grown successfully on their own for thousands of years, all humans needed to do was leave them alone and enjoy them.

As we have already learned, responsibility for administering the three new parks fell to the secretary of the interior, and he promptly called upon the War Department for assistance in policing the parks. With the exception of the Spanish-American War years, cavalry troops would come annually to the Sierra parks for the next two decades, but the problem of defining their responsibilities there would trouble military officers throughout that period. Interior Secretary Noble's pronouncements of October 1890 had provided the parks with their first regulations, among which were rules that prohibited hunting and proscribed the use of firearms for other purposes. Equally clear was a prohibition on the sale and use of alcohol.

The designation of national parks in the Sierra brought new waves of visitors to the Big Tree groves. Here, in a photo taken around 1903, tourists visit the General Grant Tree of General Grant National Park.
COURTESY OF NPS, SEQUOIA AND KINGS CANYON MUSEUM COLLECTION

Soldiers had little trouble interpreting this kind of straightforward policing and promptly went to work enforcing these policies. But what to do about the more general instructions that they were to protect "trees, shrubs, plants, timber, minerals, mineral deposits, curiosities, wonders" and so forth? Here, military officers found themselves required to exercise judgment in fields far beyond their areas of expertise. Turning in some desperation to the court of public opinion for guidance, the military soon found itself involved in two landscape-wide programs with broad implications.

These two priorities reflected the thinking of John Muir—by now a writer and public figure of some note—as well as his circle of friends and supporters. As we have seen, Muir was involved in the campaign to create a national park surrounding the Yosemite Grant lands, and his vision of what that place ought to be now carried considerable sway within the military regime designated to administer the place. From the beginning, Muir and his fellow thinkers made it clear that the protection of natural features in the Sierra would require the termination of commercial grazing in the parks. In the new Yosemite National Park this meant chasing summer herds of domestic sheep and cattle out of the High Sierra. A similar plague of grazing animals kept troopers busy in Sequoia National Park during the early years of the 1890s. Ending large-scale grazing brought immediate benefits to the parks, and within a few seasons, native vegetation sprung back with renewed vigor. Muir's other priority, however, would have a more complex effect on the forest, and especially on the Big Trees.

Muir's acute observational skills frequently provided him with ecological insights that anticipated the results of modern science. He understood intuitively the complex and interrelated character of what we now call biological systems, and he grasped a surprising number of those systems' connecting links. Yet, he was a man of his times, and one of the areas where he did not escape the thinking of his era was in his understanding of the role of wildland fire. Muir, quite simply, looked to nature for joy and inspiration and could not escape the perception that fire was a malignant and destructive agent.* This attitude had been clearly expressed in Muir's early Sierra writings, and the paradigm remained constant throughout the following decades. In February 1876, when Muir addressed the *Sacramento Record-Union*, he succinctly laid out the position he would hold for the rest of his life:

> Fire, then, is the arch-destroyer of our forests, and sequoia forests suffer most of all. The young trees are easily fired-killed; the old are most easily burned, and the prostrate trunks, which *never rot,* and would remain valuable until our tenth centennial, are reduced to ashes.[1]

As late as 1912, just two years before his death, Muir would still call fire "the great destroyer of sequoia," as he did in *The Yosemite*, his last book.[2]

On this subject, Muir represented the mainstream of his time. Hence, it will come as no surprise that the other landscape-wide initiative undertaken by the army in the early years of the national parks was the beginning of a program to suppress fires. In this task the army possessed several advantages, including manpower and organizational discipline. It also found itself working (although this point escaped most of the soldiers) in forest settings that remained in relatively low-fuel states as a result of

---

* The modern acceptance of fire as a formative and often positive part of forest systems would not come into focus until much later. See Chapter 15.

centuries of natural wildfires, as well as those set by Native Americans for ecologically sound purposes.

The native peoples of the Sierra Nevada understood that fire was not evil and was, in fact, beneficial to the health of the forests. Nearly all of California's native groups, including the Miwok and Western Mono peoples of the Sierra, kindled fires to manage vegetation and game. These fires thinned forests and enhanced many plant species that were then browsed and grazed by wildlife. The combination of these intentional fires with those ignited naturally by lightning kept the forests of the Sierra Nevada relatively open and thus resistant to intense, all-consuming fire events. This was the landscape nineteenth-century Euro-Americans found when they first moved into the Sierra, but only a few recognized it for what it was.

The fire question came to a head during the 1890s in the adjoining Yosemite National Park and state-managed Yosemite Grant lands. Beginning in 1891 with arrival of federal cavalry troops, this region witnessed two parallel and often competing management systems: one for Yosemite Valley and the Mariposa Grove, and the other for the surrounding hundreds of square miles of national park, including the Merced and Tuolumne Groves of Big Trees. As early as 1889, the fire issue had surfaced in hearings in Sacramento about the management of the Yosemite Grant by the Yosemite Park Commission. Disgruntled commissioner William H. Mills, who resigned from the commission later that year, testified that the complete suppression of fire in the Mariposa Grove was allowing an unnatural accumulation of flammable fuels and that, as a result, the sequoias were increasingly at risk. After 1891, the army wrestled with the same question, and with inconsistent results. Several of the military officers appointed to serve as acting superintendents of Yosemite National Park accepted the logic of "light burning" to reduce fuels, while officers in other years actively suppressed the same sorts of fires. In 1897, for example, troops intervened to prevent a summer fire from entering the Merced Grove of giant sequoias.[3]

By the early years of the twentieth century, the anti-fire school dominated public thought, and talk of light burning in the Sierra's forests faded away. Muir's attitudes, for once in line with broad public opinion, won out, and fire suppression became the norm in and around the

Sierra's sequoia groves. Few early decisions would have more profound long-term consequences for the forests of the Sierra Nevada.

Even as the US Army and the Yosemite Commission struggled with how to care for the giant sequoias under their two administrative systems, the fate of the Sierra's many other groves remained an open question. Together, the Yosemite Grant land and the Yosemite, Sequoia, and General Grant National Parks contained within their boundaries roughly one-third of the Sierra's Big Trees. The men who had organized and advanced the two 1890 campaigns to create the national parks understood this, and they saw the successes of that year as just an opening step in a larger campaign to bring the entire range under a program of management and protection. Again, leadership emerged from both the Tulare County group, centered around George Stewart, and the faction that included John Muir and his Bay Area friends.

With the quiet help of the Southern Pacific, Muir and his supporters had achieved substantially more in 1890 than they had set out to accomplish. The new Yosemite National Park contained some forty-two townships of thirty-six square miles each, and this extensive tract included the majority of the watersheds of the Merced and Tuolumne Rivers. All of this land had been withdrawn from sale. To the south, however, the picture was not so positive. The new Sequoia National Park was only one-sixth the size of Yosemite, and most of the rest of the southern Sierra remained for sale by the General Land Office. This huge and unmanaged domain included dozens of giant sequoia groves.

George Stewart had begun his work in this area with the broad objective of protecting the entire southern Sierra, and following the creation of Sequoia National Park, he now returned to that goal. As early as April 1891, he was editorializing in the *Visalia Delta* that the national park in Tulare County was too small and needed to be extended eastward to the Sierra Crest.[4] Meanwhile, Muir began to write from a similar point of view. In the May 1891 issue of *Century Magazine*, Muir detailed the need to create a huge national park in the southern Sierra, similar to Yosemite National Park farther north. The proposal centered on the famous glacial canyon of the South Fork of the Kings River but concluded with a call that not only this region but also the "Kaweah and Tule sequoias" should be placed within "one grand national park."[5]

These proposals, despite their heartfelt sincerity, did not catch on, but soon a new opportunity arose that had more promise. Even as Stewart and Muir renewed their efforts to designate much of the southern Sierra as a national park, Congress was at work on a bill to modify public land policies. In those days, new Congresses took office in the month of March, and in February 1891 the expiring Congress was still wrestling with proposed legislation to repeal the Timber Culture Act of 1873, yet another public land law that had facilitated large-scale fraud. Added to the bill as it finally emerged from committee hearings was Section 24, which gave the president the authority to withdraw public-domain forest lands from sale and set them aside as "forest reserves" to be held in permanent public ownership. This proposal had the support of John Noble, the same man who had during his term as secretary of the interior both supported the creation of the California national parks and arranged for the destruction of the giant sequoia tree that would go to the World's Columbian Exposition. Sending this bill on to President Harrison, Congress adjourned. On March 3, 1891, Harrison signed the bill, which gave to him and his successors a broad new executive power over forest protection.

Seeing the potential offered by this new statute, the Tulare County men interested in preserving the southern Sierra changed gears and began to work for the creation of a forest reserve. They approached their congressman, and in response the General Land Office assigned special agent B. F. Allen to the question in October 1891. Allen spent the entire summer of 1892 in the mountains, and by early 1893 he had completed his report. It called for a reserve of nearly 6,000 square miles. On February 14, 1893, President Harrison, about to leave office, signed the proclamation that Allen had drafted.[6]

With the establishment of the Sierra Forest Reserve, the federal sale of forest lands in the Sierra Nevada came to an abrupt and permanent end. For the giant sequoias, this meant that all the groves that had not yet been sold under the Timber and Stone Act or other federal statutes would remain permanently in public ownership. The government would create no more opportunities like the one that facilitated the eventual destruction of the Converse Basin, a process that was only just getting started in 1893.

At the time, the men who had lobbied for the creation of the new reserve saw little practical difference between the Sierra Forest Reserve and the three new national parks. After all, the laws that had created the parks had legally designated them as "forest reserves" in name. Special agent Allen also considered the labels interchangeable, and in his report he described this new protected area as "a great National Park" that "will attract visitors from all over the world."[7]

History, of course, would ultimately take these reserved lands in very different directions. The forest reserves would not become national parks. Instead, after a decade of neglect by the Interior Department, Congress would respond to a request from President Theodore Roosevelt and transfer the reserve lands to a new agency in the Department of Agriculture: the United States Forest Service. By 1908, the forest reserves had all been reorganized into a new system of national forests under the management of chief forester Gifford Pinchot. But that's a story for a later chapter.

How influential were the giant sequoias in the events of 1890 and 1893? An honest answer must be that the existence of the trees played a varying role in these seminal decisions. The presence of the Merced and Tuolumne Groves, for example, cannot be seen as critical to the energy thrown behind the creation of Yosemite National Park; compared to the Mariposa Grove, which remained in 1890 under state control, the two relatively small sequoia groves within the new national park were no more than minor features augmenting the main attractions: the mountains. Sequoia and General Grant National Parks, however, clearly owed their protection to the presence of the Big Trees, and Interior Secretary Noble confirmed this emphasis by naming the two new parks for their trees. General Grant National Park, covering a mere four square miles, existed only because of its sequoias. Neighboring Sequoia National Park, despite its much larger size, also focused on the Sierra's forest belt and the special trees within.

For the American national park system, 1890 was a key year. Prior to that, the nation had been content with its single national park—Yellowstone—under the assumption that it would function as *the* national park for the country as a whole. That model obviously changed in 1890. The efforts of a handful of Californians brought to life the idea of a national park *system*, and in this development, the charismatic giant sequoias played a highly significant role. The ability of the Big Trees to serve as a catalyst for political action had reached a new high.

Although the sequoias played a lesser role in the designation of the Sierra Forest Reserve in 1893, in the long run it would become apparent that the lands withdrawn from sale in that process actually contained more sequoias than Yosemite, General Grant, and Sequoia National Parks combined. No one seems to have fully understood at the time that the forest reserve proclamation was a godsend, especially for the Tule River groves.[8]

The political decisions made in 1890 and 1893 determined the ultimate status of most of the Sierra's giant sequoia groves, but the future of one property remained unresolved. In the years following 1893 this tract almost became the Sierra's fourth national park dedicated to the protection of giant sequoias; the story of how this effort failed now demands our attention.

As the Yosemite region emerged as the Sierra's leading tourist destination, the Mariposa Grove assumed the mantel of the range's must-see giant sequoia grove. Particularly after 1886, when the Southern Pacific Railroad opened a branch line to Raymond, located only twenty-five miles from Mariposa, the Wawona route to Yosemite Valley took on added importance. Trains now brought tourists to Raymond, and stagecoaches took them on to Wawona and Yosemite Valley. The Washburns' hotel at Wawona catered to this trade, and most of the tourists who enjoyed this route also took the short side trip to the Mariposa Grove with its famous tunnel tree. During these years, the facilities at Calaveras remained popular, but without easy access to the wonders of heavily marketed Yosemite Valley, the northern grove's popularity declined.

The North Grove of the Calaveras Big Trees—still the site of the Mammoth Grove Hotel that had been erected in 1856—remained in private ownership. After 1878, when a series of private-party lawsuits over title to the grove were resolved, James L. Sperry controlled both the Big Trees and the resort facilities that stood on the edge of the grove. By the early

1890s, Sperry was ready to wind down his business activities, and he began to search for a buyer. Sperry talked to state officials several times, offering them the grove at a reasonable price, but nothing came of the discussions. California attorney E. S. Pillsbury attempted to have the grove added to Yosemite National Park in 1892, and even went so far as to travel to Washington, DC, to lobby for the proposition, but this effort failed as well. Finally, in 1899, Sperry, who by now was approaching seventy years of age, found a taker.

On January 1, 1900, Sperry signed the papers that gave control of the grove and its associated improvements to Robert Whiteside, a businessman from Duluth, Minnesota. The agreement committed Whiteside to paying $100,000 for the 2,320 acres that contained the Big Trees, a sum that valued the famous forest at less than $50 per acre and threw in the hotel as a bonus. Whiteside had established his substantial fortune by pursuing mining and lumbering interests in the upper Great Lakes region, and he apparently recognized the opportunity to make a bargain purchase in California. In addition to the lands containing both the North and South Groves at Calaveras, he also moved to purchase 5,000 acres of neighboring forest. Almost immediately, rumors began to spread about his intentions, and Whiteside's motives came under close scrutiny. The directors of the Sierra Club, founded in 1892 in San Francisco with John Muir as the club's president, promptly took up the question at a special meeting. By January 23, one of those directors, Stanford University president David Starr Jordan, had taken up his pen and addressed the secretary of the interior on the matter. In his letter Jordan described the two groves, with some hyperbole perhaps, as "the finest of the great forests of the world," and he warned that their destruction would be a "national calamity." He noted the lamentable ongoing destruction of the Converse Basin as well as the successful preservation of the Mariposa Grove and the groves in Tulare County located within the two national parks there. Getting to the point, he concluded by suggesting that "nothing in the way of forest preservation can be more important than the making of the Calaveras Grove a public park."[9]

After that, things moved quickly. On February 19, California senator George Perkins introduced a joint resolution calling for the federal acquisition of the Calaveras tract. By March 8, both houses of Congress had approved the resolution. It called for the secretary of the interior to initiate negotiations with Whiteside to ensure the "preservation, management, and control" of the Calaveras Big Trees.[10] Only one obstacle remained: Mr. Whiteside, it turned out, had no interest in selling the grove unless he could turn a good profit. He had invested $100,000 to gain ownership of the property, and he now made it clear that he would not sell it for less than several times that amount. This brought everything to a halt. Congress balked, being unwilling either to pay a much higher price for the grove than had Whiteside or to undertake the legal condemnation of private land for park purposes.[11]

Despite this setback, those in favor of the new park kept working on the problem. In response to their pressure, the Senate requested that the Agriculture Department's Division of Forestry compile a report on the status of the giant sequoias as a species. The resulting document—prepared, it was later learned, by government forester George Sudworth, with the assistance of Stanford University forestry professor William R. Dudley (also a Sierra Club director)—pulled together most of what was known at the time about the sequoias in terms of both biology and geography. Large foldout maps detailed the locations of the known sequoia groves as well as the extent of private ownership within the groves.[12] Sudworth gathered material from all available sources, including information from Josiah Whitney, James Hutchings, and John Muir. Reflecting the government's perspective, the report identified important values in the Calaveras Groves, making the argument that aside from the Big Trees at Mariposa Grove, most of the other sequoia groves had been so compromised by land sales that their complete preservation could not be achieved.

All this eventually came to naught. Robert Whiteside would hold on to the Big Trees at Calaveras for the next three decades. Several additional attempts to bring them into government ownership, including a 1912 effort to create a "Calaveras Big Trees National Forest" through land exchanges, all failed.[13] Not until 1931 would Whiteside finally let go of his sequoias. The good news is that during those long decades that he controlled the Calaveras Big Trees, he focused on tourism and did not allow logging within the two groves. The Big Trees there, unmolested, grew ever larger.

Surprisingly, the failure of the campaign to create a Calaveras National Park was followed by a successful effort to bring into national park status perhaps the best known of all the groves. Since 1890, frustration had continued to fester regarding the management of the Yosemite Grant by its state commissioners. These complaints, which focused on matters as varied as poor roads and trails and an excessive willingness to defer to commercial operators, found particular resonance among the Bay Area residents who had formed the Sierra Club. The question was one of the first to be addressed after the club was founded in September 1892. The Yosemite commissioners defended their management, and over the next several years the directors of the Sierra Club made their peace with a status quo that continued to see both Yosemite Valley and the Mariposa Grove as state-controlled islands within a larger setting of federally managed national park and forest reserve lands.[14] Muir and his friends backed off, but they did not forget the issue, and circumstances eventually arose that gave them a new chance to push for the recession of the Yosemite Grant lands to federal control.

In May 1903, President Theodore Roosevelt visited California, and one of the places he wanted to see was Yosemite National Park. He also wanted to go camping with John Muir. The California naturalist, now sixty-six years old, had other plans but agreed to delay them for the priceless opportunity of spending time with the outdoors-minded president. Roosevelt and Muir arrived together at Raymond on a Southern Pacific Railroad presidential special from the Bay Area. After the president addressed the welcoming party waiting there, his group boarded two eleven-passenger stages for Wawona. They went up to into the Mariposa Grove first, even before they went to Washburn's hotel. There, among the Big Trees, Muir had his first real opportunity to talk to the young president about the natural world the older man cared about so much.[15]

At this point, under the influence of the Big Trees, matters took an unexpected turn. A large party awaited the president at the Wawona Hotel, where a banquet had been laid out, but the president failed to arrive at the hostelry as scheduled. In fact, he failed to arrive there at all. Instead, Roosevelt and Muir, in the company of two park rangers and two cavalry soldiers, abandoned the rest of their official group and spent the night

On a cold spring day in 1903, John Muir and President Theodore Roosevelt visited the Grizzly Giant Tree in Yosemite's Mariposa Grove. Soon after this photo was taken, the naturalist and the president jettisoned their official companions and spent several days alone exploring the park. COURTESY OF YOSEMITE NPS LIBRARY

camping among the Big Trees on informal mattresses made of pine boughs. For the next forty-eight hours, the president continued to play hooky with Muir, camping for a second night at Glacier Point (sleeping through a snowstorm!) and then for a third night at Bridalveil Meadow, at the western end of Yosemite Valley. All this time, the two men talked about nature and the need for conservation.[16]

Roosevelt's brief time with Muir had broad and enduring consequences, and among them was the rejuvenation of the campaign to return Yosemite Valley and the Mariposa Grove to the federal government. After a second campaign, the transfer was authorized by the State of California in early 1905. Working quietly to get the bill through the legislature was again the Southern Pacific Railroad, now under the control of Muir's friend Edward Harriman. After some confusion at the federal level having to do with proposals to change the boundaries of Yosemite National Park,

President Roosevelt signed the bill accepting the lands in question on June 11, 1906.[17] The world's best-known and most-visited sequoia grove had become a national park feature!

The last years of the nineteenth century witnessed fundamental changes in the status of the giant sequoia groves of the Sierra Nevada. Before 1890, the great majority of the Sierra's sequoia groves were for sale by the federal government at nominal prices. Only a handful of groves had found any sort of protection, and two of those situations were tenuous. Four square miles around the Grant Grove had been held back from sale since 1880, but that could be undone just as easily as it had been decreed. For decades the Calaveras Big Trees had been essentially protected by private owners who had dedicated them to tourism, but as events were about to prove, those owners could change their minds as well.

Within a few short years, however, almost everything was different. The creation of the three Sierra Nevada national parks in the fall of 1890 withdrew some two dozen groves from sale and placed them under permanent government control so that they could be both protected and enjoyed. It would take decades for the government to work out exactly what that protection and enjoyment ought to look like, but a start had been made. Three years later, the designation of the Sierra Forest Reserve withdrew from sale essentially all of the remaining public lands that contained Big Trees. In theory at least, the forest reserve would protect the two great clusters of sequoias that were not otherwise preserved in national or state parks—those along the ridges south of the main fork of the Kings River and those in the Tule River watershed.

This protection came too late, of course, for many of the Kings River trees. There, the effects of Timber and Stone Act sales had already led to sequoia logging on an unprecedented scale, and that logging would continue for several additional decades before it finally exhausted itself. Among its victims would be the Converse Basin Grove, in hindsight perhaps the Sierra's ultimate sequoia forest. In the Tule River country, the results of the creation of the forest reserve were more complicated. The region's biggest sequoia groves, particularly Mountain Home and Dillonwood, were already checkered with private holdings and small-scale sawmills, but the region's several dozen smaller and more isolated groves directly benefited from the termination of land sales in the region. These groves would be spared from the axe, at least until the federal government itself began to log within the groves several generations later.

And finally, after the turn of the century, even as Calaveras escaped government ownership to remain in private hands, the Mariposa Grove found a home in Yosemite National Park.

It would be too much to argue that the sequoias were at the center of these developments, but it's not a stretch to say that the presence of the Big Trees did play a significant role in what happened in the Sierra during this time. In a way that had no parallel in the national experience, the sequoias had become a catalyst for public and government action. Even as logging continued in some of the groves, the Big Trees of California had achieved an unprecedented status among American trees. For many, they had now become something special—objects of veneration, sacred in their own unique way.

They'd become American icons.

CHAPTER TEN
A Source of Inspiration

I look out over the polished brass railing and watch the oaks roll by. Beneath me, steel wheels click rhythmically over jointed rails. The evocative whistle of a steam locomotive drifts through the trees. I'm riding in a restored time capsule—Yosemite Valley Railroad's Observation Car 330.

I close my eyes and listen. The old car creaks like a wooden boat. As the rails curve in preparation for crossing over the stream that defines this small canyon, the car's wheels squeal in mechanical protest. I open my eyes again and gaze upward at the carefully crafted mahogany ceiling that shades this classic open-sided observation platform. We rumble across a truss-work bridge; its steel beams amplify the sounds of the train, a combination of vibrations reflecting the passage of several hundred tons of rolling iron. I let my mind drift to others who have sat here—some famous, most not. When this car was new, it hosted travelers steeped in Edwardian propriety, and I imagine men in wool suits and women in ankle-length skirts and shirtwaists, even on hot afternoons. No less than President William Henry Taft once took in the scenery from this platform, as did countless of his nation's citizens. All were headed to a single destination, an entrance to a fascinating world of towering granite cliffs, exquisite waterfalls, and unimaginably giant trees. The railroad,

Having just disembarked from their train at El Portal, early-twentieth-century visitors wait to board stages that will take them to Yosemite Valley and the Mariposa Grove. COURTESY OF YOSEMITE NPS LIBRARY

recognizing this sense of arrival, named its terminus "El Portal" when the line opened in 1907.

My reveries aside, I am not headed to Yosemite today. Train service to that destination ended in 1945, and Observation Car 330 ended up as a diner in the far northern California town of Yreka. A group of rail fans rescued the car's stripped hulk in the 1990s, moved it to the Bay Area, and began its painstaking restoration. Thousands of work hours later, the car again rides the rails, this time at the Niles Canyon Railroad, a museum in the hills southeast of San Francisco Bay. Like the Yosemite Valley Railroad of long ago, this line also follows a sinuous watercourse through oak-studded hills, but Niles Creek is a poor substitute for the Merced River, with its spirited spring snowmelt floods. Still, if one allows one's senses free reign to dream just a bit, it is possible to gain here a sense of what early-twentieth-century tourism felt like. That's why I have come—to capture something of that now-remote era when Sierra Nevada tourism moved to a new level of accessibility and comfort, and the Big Trees played muse to a generation of artists, writers, scientists, and, yes, recreational tourists.

In the early years of the new century, the growing world of Big Tree tourism led to new efforts to define both the charms and the significance of the sequoias. If the trees were to play an ever-growing role in the commercial and political worlds, then their cultural symbolism would need to be enhanced. The task fell initially to the major railroad companies.

Premier among these companies in California at the time was the Southern Pacific, based in San Francisco and running its passenger trains as far east as New Orleans and, with the help of the Union Pacific and the Chicago and Northwestern, also to Chicago. The "Espee," as the line was known, stepped up its marketing of the sequoias in 1900, when it published the first of what would be a long series of free publications about seeing the Big Trees.[1] The effort continued, and even the San Francisco earthquake and fire of 1906, which destroyed the railroad's general headquarters, failed to derail the campaign. By May 1907, the Southern Pacific's list of free tourist booklets included not only *Yosemite Valley and the Mariposa Grove* but also *Big Trees of California* and *Kings and Kern Canyons and the Giant Forest*.[2] The Passenger Department of the Southern Pacific kept the *Big Trees* booklet in print for years, and it grew over time to a full thirty-two pages.[3] Comprised mostly of full-page photographs, the booklet gave equal billing to the groves at Calaveras, Yosemite, and the southern Sierra national parks. The booklet featuring the Kings and Kern Canyons provided detailed instructions on how to visit Sequoia and General Grant National Parks and especially the Giant Forest, where, "in the midst of the forest, one meets a most cordial hospitality; cots in tents kept scrupulously clean by a tidy housekeeper, and good meals served in a pleasant dining-room."[4]

In its *Big Trees* booklet, the railroad waxed eloquently about the sequoias and their apparent age:

> These trees were living towers when Cheops dreamed of building pyramids, and the hosts of Pharaoh perished in the Red Sea. Three score and ten years make up the life of man—seventy times seventy years but bring the Sequoia gigantea to a ripe maturity.[5]

On other pages, less poetic comparisons surfaced:

> Enough lumber here to make one telegraph pole forty miles high or to supply a line of poles from Kansas City to Chicago.[6]

And the Southern Pacific was not alone in this marketing. By 1901, central California had been invaded by a competing company, the Atchison, Topeka, and Santa Fe Railway, which operated transcontinental trains through the San Joaquin Valley. Santa Fe trains ran east all the way to Chicago, and that company, too, found it useful to publicize tourist destinations in California, including the Big Tree national parks of the Sierra Nevada.[7]

Most of the information in these publications derived from the work of a small, exclusive cohort of original sources. To anyone who knew the Big Trees, most of the names were familiar. In 1907, for example, Galen Clark published *The Big Trees of California: Their History and Characteristics*.[8] More than a half century had passed since Clark had first stumbled upon the Mariposa Grove and then opened his pioneer hotel at what became Wawona. By the early years of the twentieth century, Clark, now approaching ninety years of age, had become a revered and respected Yosemite figure. For years he had served the Yosemite Commission as guardian of the Yosemite Grant lands, and after his retirement from that post for a second time in 1896, he had settled again into a comfortable role as a guide and local character. Money remained a problem for the old man, however, and friends eventually talked him into publishing some of his many stories so that they could be sold to tourists. His first effort, *Indians of Yosemite Valley and Vicinity: Their History, Customs, and Traditions*, came out in 1904.[9] The volume captured many of the stories Clark had collected during his early days in the Yosemite region in the 1850s and 1860s, but it had little to say about the Big Trees. The 1907 book filled that gap.

By the time *The Big Trees of California* came out, Clark had reached the age of ninety-three, and he made no attempt to express himself at length. Instead, the old woodsman shared the essence of what he knew about sequoias in a dozen brief chapters, several of them only a few paragraphs long. A modest man, Clark summarized the Euro-American discovery of the trees without emphasizing his own role in that process. The book then proceeded to the geographical distribution and biological characteristics of the trees. Overall, Clark got it mostly right. He noted that sequoia seeds sprouted successfully only on bare mineral soil, and he documented the presence of large amounts of tannin in the trees' heartwood and cones. He even called out and speculated on the role played by tannin-rich cone pigment, a subject that would evade full explanation for another sixty years. Going on, he explored the subject of the age of the largest sequoias, and, like nearly everyone else at the time, he made the basic error of calculating backward toward the center of the tree, using as his basis the very small annual growth rings displayed by very large sequoias in their outermost wood. Applying this logic to the Mariposa Grove's Grizzly Giant Tree, he shared the mistaken assumption that dominated most other accounts at the time:

> According to the best estimates made by the examination of the annual ring growths of old fallen Sequoias, the Grizzly Giant must be not less than six thousand years old, yet still living, grizzled with age, defying old Time with his legions of furies which have shattered its royal crown, stripped its body nearly bare, and cut off its main source of nutriment. Dying for centuries, yet still standing at bay, it is probably not only the oldest tree but also the oldest living thing on earth.[10]

Clark had also spent enough time among the Big Trees to pick up a story that was gaining currency by the end of the nineteenth century. More than a few visitors had noticed the phonetic similarity between the botanical name *Sequoia* and Sequoyah, the name of the Cherokee Indian who had developed a written form of his people's language in the 1810s. Despite the lack of any firm evidence connecting the two, speculation grew that the sequoia trees were somehow named after Sequoyah. Clark repeated the story, albeit with a subtle suggestion of doubt.

> The name *Sequoia* is supposed to be derived from Sequoia (or Sequoyah), a Cherokee Indian of mixed blood, who invented an alphabet and written language for his tribe.[11]

Over time, this unsubstantiated speculation would come to be accepted by many as undoubted truth.

Still spending his summers in Yosemite, Clark lived on until 1910, when he passed away at the age of ninety-six. He was buried in Yosemite Valley beneath a small grove of sequoias he had planted in anticipation of the time he would eventually spend with them.[12] Clark's decades-long relationship with the trees he loved so deeply might well be summed up in the poetical prologue he wrote for his 1907 book:

> And I've been to the groves of Sequoia Big Trees,
> Where beauty and grandeur combine,
> Grand Temples of Nature for worship and ease,
> Enchanting, inspiring, sublime.[13]

Clark's long-time friend John Muir outlived his Yosemite companion by four years, and the naturalist had much to say about the sequoias in his last twenty years. By the early 1890s, Muir had shifted his attention to book-length manuscripts about the natural world he so deeply revered; his first book, *The Mountains of California*, came out in 1894 and was followed by *Our National Parks* in 1901 and *The Yosemite* in 1912. All three books spoke at length about the sequoias.

Unlike Clark, Muir had early committed himself not only to structured scientific inquiry but also to the painful preparation of elaborate and extended prose.[14] Muir pursued much broader interests than Clark, and as a result he actually had considerably less firsthand experience with the trees than did his fellow writer. With only minor exceptions, Muir based all his book-length writings about the sequoias on his earlier writings about the trees. In essence, this meant that what his books had to say about sequoias was based on a small handful of sources: Muir's personal journals, the newspaper and magazine articles he had written during and shortly after his 1875 sequoia expedition, and his 1877 giant sequoia paper published by the American Association for the Advancement of Science. In his three books, Muir worked and reworked this material, each time managing to make it fresh.

Muir based his first book, *The Mountains of California*, primarily on the newspaper and magazine articles he had published between 1875 and 1882. Despite its broad title, the book's seventeen chapters focused almost entirely on the Sierra Nevada, where Muir had directed most of his attention in those years. The sequoias formed a major part of Chapter 8, "The Forests." Significant portions of this material came from his 1877 paper for the American Association for the Advancement of Science, and most of the rest was from the popular account of his 1875 sequoia trip, based on his journals, that he published in the November 1878 edition of *Harper's New Monthly Magazine* as "The New Sequoia Forests of California."[15]

In the 1894 first edition, this material runs for over twenty pages, but Muir really had little new to add to the ideas presented in the original 1870s versions. Instead, he reworked the materials largely as an editor, massaging the earlier text and seeking to weave it together into a more coherent statement. He started by describing the trees' natural range and physical attributes. He then wrestled with the potential age of the giants—citing the ring count of 4,000 years he had made during his 1875 adventure, but cautioning that most individual trees were considerably younger.[16] He next moved into questions about the trees' historical range and potential future, which he had also explored in his 1878 scientific paper. He concluded with a summary of perceived threats to the sequoias, primarily logging and fire.[17]

Seven years later, in November 1901, Muir published his second major book, *Our National Parks*. The new volume brought together ten articles that Muir had drafted for the *Atlantic Monthly* beginning in 1898. The giant sequoia material appeared in the magazine in a September 1901 essay titled "Hunting Big Redwoods." Again, the naturalist was drawing on his 1878 article for *Harper's*, but this time he produced a much more complete account of his 1875 sequoia adventure. This was the last of the ten *Atlantic* articles to be published, and it was followed quickly by the book, which came out only two months later.[18]

In *Our National Parks*, the giant sequoia material stands as a full and separate chapter titled "The Sequoia and General Grant National Parks." The title is a bit of a misnomer, for the chapter is much more about the sequoias as a species than it is about the two national parks as places. He began strongly:

> The Big Tree (*Sequoia gigantea*) is Nature's forest masterpiece, and, so far as I know, the greatest of living things. It belongs to an ancient stock, as its remains in old rocks show, and has a strange air of other days about it, a thoroughbred look inherited from the long ago—the auld lang syne of trees.[19]

After a prolonged description of the trees and an age/size chart he had compiled by counting rings on stumps, he moved into a detailed account of his 1875 giant sequoia expedition. He picked up the story as he left the

Mariposa Grove and followed it all the way to the southern end of the Big Tree range, fifty miles south of Sequoia National Park. Reflecting his skill as a writer and his ability to go back into the journals he kept during the trip, Muir filled the narrative with first-person detail. Readers learned about the travails of Brownie the mule and Muir's encounter with cattleman Hale Tharp (or more likely his hired hand James Wolverton) in the Giant Forest.[20] The story is considerably richer than in earlier versions. Among the details that emerged was Muir's claim that he named the Giant Forest during his visit, a tidbit not offered in earlier accounts of the trip and hard to evaluate now.[21]

The writing contains some of Muir's most significant giant sequoia prose, including his account of reading "over four thousand [annual growth rings] without difficulty or doubt" and his oft-quoted description of watching a forest fire burn through the sequoias in the southern part of the Kaweah watershed.[22] For sheer exuberance, it is hard to exceed Muir's carefully reworked narratives. Describing his first entrance into the Giant Forest, he wrote:

> When I entered this sublime wilderness, the day was nearly done, the trees, with rosy, glowing countenances, seemed to be hushed and thoughtful, as if waiting in conscious religious dependence on the sun, and one naturally walked softly and awe-stricken among them. I wandered on, meeting nobler trees where all are noble, subdued in the general calm, as if in some vast hall pervaded by the deepest sanctities and solemnities that sway human souls.[23]

And a few pages later, describing a meadow in the Giant Forest:

> There lay the grassy, flowery lawn, three fourths of a mile long, smoothly outspread, basking in mellow autumn light, colored brown and yellow and purple, streaked with lines of green along the streams, and ruffled here and there with patches of ledum and scarlet vaccinium. Around the margin, there is first a fringe of azalea and willow bushes, colored orange-yellow, enlivened with vivid dashes of red cornel, as if painted. Then up spring the mighty walls of verdure three hundred feet high, the brown fluted pillars so thick and tall and strong they

seem fit to uphold the sky; the dense foliage, swelling forward in rounded bosses on the upper half, variously shaded and tinted, that of the young trees dark green, of the old yellowish. An aged lightning-smitten patriarch standing a little forward beyond the general line, with knotty arms outspread, was covered with gray and yellow lichens and surrounded by a group of saplings whose slender spires seemed to lack not a single leaf or spray in their wondrous perfection. Such was the Kaweah meadow picture that golden afternoon, and as I gazed every color seemed to deepen and glow as if the progress of the fresh sun-work were visible from hour to hour, while every tree seemed religious and conscious of the presence of God. A freeman revels in a scene like this and time goes by unmeasured.[24]

Text this rich would prove hard for anyone to top, and even Muir shied from that challenge when he published *The Yosemite* in 1912. He produced the book in response to the pleas of friends to write something that could serve as a guidebook to the increasingly popular national park, but it contained much previously published material, including whole sections taken verbatim from *The Mountains of California* and *Our National Parks*.[25] The chapter on the Big Trees came largely from *Mountains*.

Master writers such as Muir and Clark described the sequoias so fully and defined their emotional impact with such clarity that most lesser authors simply deferred to them. Among them was J. Smeaton Chase, an Englishman who came to California in 1890 and established a connection with the Golden State that would keep him there until his death in 1923. Chase worked at many things during his thirty-three years in California, but he is best remembered today as the author of three first-person narratives about exploring the state on horseback.[26] The first of these, *Yosemite Trails*, came out in 1911 from Houghton Mifflin, which also published Muir.[27] The book recounted several horseback trips through the front- and backcountry of Yosemite, including a chapter dedicated to the sequoias. Although well informed about California plants and their biology, Chase spent much of the chapter pursing the aesthetics of the sequoias. Taking his blankets into the woods, he described a night spent at the base of the Grizzly Giant:

> I felt much as one might who, walking among the grey ruins of Babylon or Thebes, should come upon some primeval man, ancient as the very earth, who, overlooked by death, had lived on from age to age, and might now live to the last day of Time. Its great arms were uplifted as if in serene adoration, and all around, the lesser forest stood aloof, like the worshippers in an outer temple-court, while this, their high-priest, communed alone.[28]

In other sections, Chase demonstrated sensitivity to how the trees were being managed:

> I cannot understand how...the childish, unsightly, and paltry practice could have arisen, and could continue apparently without objection, of labelling them with the names of cities, states, and persons. I confess I am amazed at the general obliviousness to the disgrace of the thing, even among cultivated persons, and am compelled to believe that the people who come to view them have no real appreciation of their grandeur, but look upon them merely with a Barnum eye as curiosities and "big things." Their admiration for the Sequoias seems to be of a commonplace and commercial kind, for there is no recognition of the anomaly involved in disfiguring objects of such nobility and beauty with hideous tin labels.[29]

As the new century began, writers were not the only creative souls trying to capture the Big Trees of the Sierra Nevada in a way the public could grasp. Graphic artists, particularly painters and commercial photographers, also sought to portray the feeling of the sequoia forests and their reigning monarchs. Of these, none was better known than artist Thomas Hill.

Born in England in 1829, Hill had come to the United States as a teenager with his father and begun studying art in his twenties in Philadelphia. He came to California in 1861, established himself in San Francisco, and was soon painting Sierra Nevada scenes in a Hudson River School style. Hill eventually became known for his paintings of Yosemite Valley, and he maintained an artistic relationship with the Sierra for the remainder of his long life. He wrestled with making a living within the endlessly changing tastes of the art market, however, and seeking financial stability, he opened a studio in Yosemite Valley. When a falling tree destroyed it, he moved to Wawona, where he found a congenial home. In 1885, Hill's daughter Estella married John Washburn, a key member of the family that owned and ran the Wawona Hotel. The next year, the Washburns erected a Victorian-style studio for Hill immediately adjacent to their hotel. Hill stocked the studio not only with his paintings but also with natural curios, including animal skins, pinecones, and Native American baskets. Over the next decade, Hill turned out hundreds of usually small paintings that he offered for sale to hotel patrons at Wawona. Many of these featured scenes of the nearby Mariposa Grove. Much of what Hill painted for his Wawona gallery did only faint justice to his true talent, but it paid the bills. Some years his income from the studio exceeded $8,000, a very substantial sum in the late nineteenth century.[30]

Hill did not entirely abandon more ambitious projects, however, and for the 1893 World's Columbian Exposition he prepared a large triptych that featured the Big Trees of the Mariposa Grove. Portraits of the Grizzly Giant and the Wawona Tunnel Tree flanked an even larger view of the sequoia forest in all its glory. On display in the art gallery that was a part of the fair's California Building, the paintings were among the first to show the sequoias not as solitary forest monarchs but rather as objects of human affection, with tourists standing at their bases. Over the next few years, Hill would paint numerous copies of these paintings, many of them on planed slabs of sequoia wood.[31]

Hill spent the rest of his life working out of Wawona. His painting activity dropped off after he suffered a stroke in 1896, but he maintained his Wawona studio and his niche as a public fixture. When Theodore Roosevelt visited Wawona in 1903 with John Muir, he made a point of meeting the aging painter, who gave the president one of his paintings. Hill died in 1908 at his winter home in Raymond, just a few dozen miles from the Big Trees he had painted so often.[32]

Sublime. Grand. Noble. Awe-inspiring. The giant sequoia literature and art of the early twentieth century defined the cultural norms that had come to be associated with the Big Trees. The sequoias deserved not just to be respected in this view but also almost worshipped, especially as these same attitudes often led to other human actions, including increased protection for this singular natural wonder.

After 1890, when Sequoia, Yosemite, and General Grant were added to Yellowstone to create a national park *system* in the United States, the list of parks continued to grow. By 1910, the federal parks roster included Mt. Rainier, Crater Lake, Mesa Verde, and Glacier. Although described even then as a system, in fact these units were anything but; each park superintendent reported independently to the secretary of the interior, and there was no such thing as unifying policy or direction. Out of this confusion came a campaign to give the parks unitary management in the form of a federal bureau of national parks. Interior Secretary Richard Ballinger recommended such an agency in 1910, but Congress chose not to act. His successor, Walter Fisher, also endorsed the idea but failed to move the issue forward. That was how the matter stood when Woodrow Wilson assumed the presidency in March 1913.

Wilson appointed Californian Franklin Lane as his Interior Secretary, and Lane brought University of California professor Adolph Miller to Washington to coordinate national park affairs for the secretary's office. Soon thereafter, when Miller was tapped by President Wilson to join the just-created Federal Reserve Board, the Interior Secretary found himself again in search of someone to lead his national park program. Lane now turned to fellow University of California graduate Stephen T. Mather, a charismatic, sometimes manic, and extremely wealthy businessman living in Chicago.

Although born in New England, Mather had deep ties to the West. He thought of himself as a Californian and had an abiding interest in California affairs. He had joined the Sierra Club in 1904, taking part in at least two of its month-long annual outings, and represented the club informally in Washington during congressional hearings regarding the proposal to build a dam in Yosemite's Hetch Hetchy Valley. In July 1912, Mather joined an extended Sierra Club outing into the backcountry of the southern Sierra and, by luck, after it ended, encountered John Muir, who was making what would be his last visit to the Giant Forest. During 1913, Mather corresponded extensively with both his Sierra Club friends and

Interior Department staff about the need to enlarge Yosemite and Sequoia National Parks with the goal of protecting more of the High Sierra.

In December 1914, Lane invited Mather to Washington and talked to him about taking on the national parks bureau project. To assist Mather, Lane offered him Miller's young assistant, a law student named Horace Albright.[33] Mather returned to Washington in January 1915 and promised Lane a year's work on the national park bureau project. As it turned out, Mather would work for the parks for the next fifteen years.

Together, Mather and his young assistant set out to convince Congress to act on the need for a national park bureau. Mather had made his mark on the world, as well as his fortune, through his work in public relations; in the jargon of a later age, he thought like a "mad man." As the spring of 1915 passed, Mather initiated a nationwide campaign to publicize the national parks and their needs, and a national parks conference in San Francisco in March was followed by politicking back and forth between the two coasts. In the meantime, Mather and Albright were plotting what they hoped would be a grand political gesture: a luxury tour of the wilderness of the southern Sierra.

Mather knew how strongly his 1912 trip through the same region had inspired his own commitment to national parks, and he now hatched a scheme to pull together a group of highly influential men and take them into the same wilderness. By the time the group emerged, he hoped, its members would be committed to supporting his national parks bureau campaign. The party gathered in Visalia in mid-July. In addition to Mather and Albright, it included Congressman Frederick Gillett of Massachusetts, Gilbert Grosvenor of the National Geographic Society, Henry Fairfield Osborn of the American Museum of Natural History in New York, and a number of writers and newspapermen. Government officials included the state engineer for California and the chief geographer of the United States Geological Survey. The Southern Pacific Railroad, always interested in national park affairs, sent its vice president, Ernest O. McCormick.[34]

Mather knew how he wanted this trip to begin, and that is where the sequoias came in. The morning after the opening banquet in Visalia, the group motored to the Giant Forest and set up its first camp among the Big Trees. There, as they got to know each other better, Mather's carefully chosen party dined and slept *al fresco* beneath the giant trees. An elaborate dinner, complete with white linens, was set out on a rough table made of logs. Under the magic spell of the Big Trees, the group began to coalesce.[35]

The "Mather Mountain Party," as the group came to be remembered, went on to spend ten days in the backcountry of the southern Sierra and became the central core of Mather's national park campaign supporters. Thirteen months later, just as Mather was coming off a second High Sierra trip, he received word that President Wilson had signed a bill to create the National Park Service.

An enduring link had been forged between the new agency and the Big Trees. The first official emblem of the National Park Service would feature a giant sequoia cone as its central motif and, by the 1930s, all park ranger uniform belts and leather hatbands were embossed with a sequoia cone pattern. In 1952, when the agency updated its imagery, the new "arrowhead" emblem featured the silhouette of a giant sequoia tree. Even today, a century after the creation of the National Park Service, all park rangers still wear uniforms decorated with images of California's giant sequoias.

CHAPTER ELEVEN
Science and Time

Getting here has been a challenge. I've spent several hours coaxing my old 4x4 eastward along a hundred-year-old logging railroad grade into one of the least visited of all the Big Tree groves that one can attempt to drive to. If my destination hadn't been so distant, I would have done better to walk. For decades the Forest Service has done little to keep this route passable, and rocks too large to roll out of the way choke some of the old railroad cuts. Getting around them requires a slow-motion, sideways form of vehicular limbo as I wend my way ever so carefully between and sometimes over the angular chunks of granite. Like many another 4x4 driver, I have pushed farther than I should have, but I keep going, since I've now invested so much into getting this far. Eventually, to my relief, the route improves again; the worst is now behind me, or so I hope.

I've been skirting along the precipitous south rim of the great canyon of the Kings River. The Hume-Bennett Lumber Company built this railway in 1914 with the goal of extending its logging operations into the untouched sequoia groves to the east of Converse Basin, along the rim of the Kings River Canyon. The Hume-Bennett Company knew a good deal about logging giant sequoias because it was the successor to the Sanger Lumber Company, the corporation that had cut the Big Trees of Converse Basin. When that grove was at last exhausted, the company moved eastward a few miles and relocated its sawmill in 1909 to the shores of newly constructed Hume Lake. Thomas Hume of Michigan had purchased the business in 1905 knowing that although Converse Basin's time was nearly over, the company also owned, thanks to the Timber and Stone Act, large tracts of untouched land to the east. These forests included the mountainside sequoia groves along the steep rim of the Kings River Canyon.[1]

Now, a century after the Hume-Bennett operations moved into these groves, I am tracing their old rail line. It's easy to tell that this sequoia forest, now called Evans Grove, was harvested. During the logging seasons of 1914 through 1918, millions of board feet of freshly cut timber rolled down this rail line to the Hume Lake sawmill, much of it consisting of giant sequoia logs. Accurate figures are elusive, but the best estimates suggest that several thousand sequoias were cut here, including perhaps a thousand truly large trees.[2] Today, massive sequoia stumps dot this mostly regrown forest, making it clear that the pines, firs, and young sequoias that dominate here now have a long, long way to go to match the forest that stood here before the lumbermen arrived. I've come to study these stumps, and eventually I find one that documents the story I seek.

Like many others, this stump rises some five or six feet into the air, then ends abruptly in the sawn surface left by the loggers who felled the tree. What distinguishes this stump, however, is that cut into the top of this now-weathered surface is a long V-shaped notch perhaps six inches wide and deep. The incision, clearly made by a long saw, runs cleanly from the center of the stump to its perimeter. I brush accumulated pine needles out of the notch and run my fingers over the hundreds of annual growth rings exposed in its sides. I know who did this. In August 1915 a young astronomer from the University of Arizona came here to study tree rings. Working on the stumps of freshly fallen trees, he cut out what he called "radial sections" and took them back to Tucson to study in his lab at the university. He didn't know yet just where his research would eventually lead, but A. E. Douglass was already well on his way to inventing the science of dendrochronology.

The story begins, however, not with Douglass but rather with Ellsworth Huntington, a highly confident, almost egocentric professor of geography at Yale University. Time has not been kind to Huntington's reputation, which is marred by his later endorsements of eugenics, but in the early years of the twentieth century he was both a prolific and an influential scholar.* Born in Illinois in the centennial year of 1876, Huntington spent much of his twenties in the Near East and Asia. He taught for four years (1897–1901) at Euphrates College in what is now Turkey, and then took part in several expeditions into central Asia. He joined the staff of Yale University as a geographer in 1907, and in 1909 led that institution's expedition to Palestine. Out of these formative experiences came theories that would be central to Huntington's professional life as a geographer.

In the Near East and Asia, Huntington was struck by two things: the question of why civilizations collapse, and the fact that much of the region showed historical evidence of increasing aridity. Putting these two together, Huntington concluded that climate change was a likely cause of the rise and fall of civilizations. He initially expounded this idea in a 1907 book titled *The Pulse of Asia: A Journey in Central Asia Illustrating the Geographic Basis of History*.[3] Continuing his studies, Huntington shifted his attention to the American Southwest, another arid environment with evidence of failed civilizations. It was in this setting that Huntington discovered the pioneering work of A. E. Douglass.[4]

Andrew Ellicott Douglass had come to the University of Arizona in 1906 as an astronomer after spending twelve years working with astronomy pioneer Percival Lowell in northern Arizona. During his work with Lowell, Douglass had become interested in astronomical cycles, particularly the eleven-year sunspot cycle, and he had hit upon the idea while living in Flagstaff that tree rings in dry climates might preserve evidence of these cycles. In 1909, Douglass published a preliminary paper in the *Monthly Weather Review* proposing just that.[5] This paper came to the attention of Huntington, who immediately recognized that Douglass might have

---

* Huntington's interest in eugenics led him to espouse racial theories that most find repugnant today.

stumbled upon a means of documenting climate change over the past several thousand years.

If tree rings offered a possible window into past climates, then, at least in theory, the best record would come from the oldest trees. This is what brought Huntington to California in the summer of 1911 to study giant sequoia stumps. The antiquity of the Big Trees was by then both well known and often exaggerated. As we have seen, Muir believed that he had counted 4,000 rings on a single burned snag, and even as cautious an observer as Galen Clark had speculated only a few years earlier that some trees might be as old as 6,000 years.*6 Into this picture marched Ellsworth Huntington.

Huntington stepped off the train in Sanger, California, in late May of 1911, and together with two assistants, he headed into the Sierra with the goal of studying giant sequoia growth rings. The best place to do such work, obviously, would be an area where sequoias had been recently felled, and this is what brought Huntington to Sanger, the railhead town for the Hume-Bennett Lumber Company. Traveling by motorcar, Huntington climbed into the mountain country north and east of General Grant National Park. Completing his journey into the mountains via the lumber company's rail line, Huntington established camp in the area where lumbermen were actively cutting Big Trees—a forest area known to the Hume-Bennett people as Camp Six.7

Huntington's technique exemplified simplicity itself. Since his hypothesis was that the growth rings of the sequoias would vary over time in response to precipitation cycles, the geographer set out to collect data from numerous stumps that documented growth rates over time. Huntington built his work around the hypothesis that the trees would grow faster, and thus lay down thicker growth rings, in wet years as compared to drier growth seasons. To apply this logic, Huntington's team simply counted the number of growth rings per inch. He and his assistants wrestled with the dull work in the hot early-summer sun, fighting off periodic attacks by pugnacious

---

* Modern efforts to document the age of the sequoias have demonstrated that very few Big Trees surpass 3,000 years of age, and then only by a century or two at most.

carpenter ants. Eventually, during twelve weeks of tree-ring counting over the summers of 1911 and 1912 on both the Hume-Bennett land and in the vicinity of the Enterprise Mill in the Mountain Home Grove in the Tule River country, Huntington counted the rings on about 450 giant sequoia trees.[8] No one to date had ever studied sequoia age and growth rates in such depth, and Huntington's work provided new perspectives on the Big Trees. His main discovery was this: "It is easy to obtain plenty of young trees under 2,000 years of age, but older ones are…scarce."[9] In his 450 tree-ring counts, Huntington found only three trees that exceeded 3,000 years of age, and none of these three exceeded that total by more than a few years. The Big Trees, it seemed, were not as old as most had assumed.[10]

None of this, however, prevented Huntington from finding what he wanted to find. He had set out to demonstrate that there were worldwide patterns of climate change that could be used to explain history, and once his data was in hand, he proceeded to attempt to do just that. In an article published in the July 1912 issue of *Harper's Magazine,* Huntington produced a graph that not only purported to trace long cycles of wet and dry weather as documented in giant sequoia growth rings, but that also compared that pattern to a record of wet and dry weather cycles he had discerned from historical records in Asia. In truth, the two lines, coming from very different sorts of data, offered little to support his hypothesis, but he was not particularly discouraged. As he concluded:

> In essentials the two [lines] agree in spite of differences in detail. It therefore seems probable not only that climatic pulsations have taken place on a large scale during historic times, but that on the whole the more important changes have occurred at the same time all around the world, at least in the portion of the north temperate zone lying from 30° to 40° north of the Equator. This, in itself, does not prove that great historic changes have occurred in response to climatic pulsations, but it goes far in that direction.[11]

In short, Huntington had convinced himself, using data collected from his giant sequoia studies, that he had found a new way to understand human history. The Department of the Interior thought enough of

Huntington's article to have it reprinted and distributed to national park visitors in 1913 as a government publication. The man who had started Huntington down the giant sequoia path, on the other hand, was not so impressed.

Douglass stayed in touch with Huntington and even contributed a chapter on tree-ring analysis to the geographer's 1914 book, *The Climatic Factor as Illustrated in Arid America*.[12] But Douglass had his doubts. The Arizona scientist possessed enormous patience and a determination to be as accurate as possible. As a result, he did not fully trust Huntington's work and suspected that more careful study of the tree-ring record preserved within the giant sequoias would disclose much of significance that had simply not been of interest to Huntington. In August 1915, Douglass set out from Tucson to revisit Huntington's California study sites and take a more careful look at the tree-ring record that had been exposed by the Hume-Bennett lumbermen.

Like Huntington, Douglass was out to prove something—in this case that the eleven-year sunspot cycle affected rainfall in a way that was recorded in the growth rings of certain trees. But this question took Douglass in quite different directions than had been explored by his geographer predecessor. The most immediate difference between the two approaches was that because Douglass, unlike Huntington, was seeking evidence regarding a cycle with a short interval, he needed to identify the exact year in which tree rings were laid down. This gave his work a precision entirely missing from Huntington's, which had been concerned with long-term patterns. The astronomer asked whether all sequoias were likely to record differences in annual growth rates or only those trees in drier, more stressful locations. He tested this question by comparing growth rates of trees in wetter and drier sites and discovered that there was indeed a measurable difference. Trees growing on relatively dry sites recorded ring variation that was similar to Big Trees growing on other dry sites; trees in more favorable locations did not produce similar results. In this, Douglass confirmed what would become one of the basic tenets of dendrochronology: that tree-rings best recorded annual changes in climate when the trees in question were stressed. He called these stressed trees "sensitive" and the trees growing in wetter sites "complacent."[13]

The severely logged Big Tree forests south of the Kings River provided a perfect setting for the tree-ring research of A. E. Douglass.
COURTESY OF LABORATORY OF TREE-RING RESEARCH, UNIVERSITY OF ARIZONA

In his determination to date individual tree rings accurately, Douglass also did something that Huntington had not pursued: he cut study sections from each of the trees he worked on and took them back to Tucson for further reference. This greatly reduced the number of trees he could study in the field, but it produced much more precise results. Not only was he eventually able to determine the exact age of each ring in these samples, but he also discovered how far off some of Huntington's dates had been—sometimes hundreds of years. Because of this exacting work, Douglass could now begin to discern much stronger patterns. Here, he realized, was actual data about growing conditions hundreds or even thousands of years ago.[14]

Using data collected during his 1915 visit to the Hume-Bennett lands, Douglass succeeded in establishing a chronology of good and bad (wetter and drier) growth years in the giant sequoia groves that stretched back about 2,200 years. Hoping to extend this chronology back to 3,000 years or more, Douglass returned to Camp Six in the summer of 1918. Huntington had reported a tree supposedly aged over 3,000 years at that site, but Douglass could not find a stump with that many rings. Still in search of older trees, he extended his studies beyond the Hume-Bennett lands, and finally, on the western edge of Converse Basin, he found a stump that careful counting proved was at least 3,197 years old. On that same trip, he visited the General Grant Tree and the adjoining (and by then badly burned) Centennial Stump,[15] which he estimated to be about 1,800 years old. He continued to the Enterprise Mill site in the Mountain Home Grove, where he re-located two of Huntington's oldest stumps, both more than 3,000 years old.[16]

Douglass would continue to work on the giant sequoias. He returned to the groves in 1919, 1924, and 1925.[17] He also did similar work in numerous other forests. Over time, he came to understand what he had discovered. Old trees, as it turned out, did not document how the eleven-year sunspot cycle affected global weather, but they did offer something else, something much more specific and very useful. Carefully selected "sensitive" trees, when studied closely, could provide a window into localized climate history. By studying, cross-dating, and then analyzing tree rings, it was possible to gain insights into regional shifts in precipitation over the previous several thousand years. A. E. Douglass had invented the science of dendrochronology.

Huntington and Douglass pioneered a new way of looking at the Big Trees. Before they began counting giant sequoia tree rings, scientists had focused on the trees simply to understand their basics—mostly where and how did they grow? Now, for the first time, the trees had been approached in another way: as organisms that could reveal not just their individual stories but also information about much bigger issues—things like the history of the earth's climate. This new appreciation of the Big Trees would grow slowly, for many decades remaining limited to communities of scientists. But eventually dendrochronology would lead to a broad public awareness that the trees possessed significances that went far beyond their abilities to generate awe and attract tourist dollars.

CHAPTER TWELVE
Running into Limits

In each of our lives there are anniversaries that are uniquely our own, days on which we quietly celebrate the significant events that have made us who we are. Today, June 15, is such a day for me—a private holiday, if you will—for it was on a June 15 nearly half a century ago that fate delivered me into the world of the Big Trees.

Two weeks out of high school, I needed a summer job, so when the phone call came telling me to report to the personnel office at the Giant Forest Lodge in Sequoia National Park, I eagerly packed my duffle and took off for the mountains. When I arrived, the lodge's personnel director, a recently retired Marine Corps colonel whom I just as easily in those years could have met in Vietnam, quickly sized me up and offered me work hauling visitor luggage to the lodge's many scattered cabins. So began my life among the sequoias, a continuing adventure that would fill the summers of my college years.

Now, in a quiet moment of remembrance, I'm back where that story began. Around me tower the individual sequoias I came to know in those years. They are old friends, the best of them as familiar as family. One leans over a place where I used to try not to park my car. I didn't trust that tree, but it still stands tall, even if it diverges almost ten degrees from vertical. Another tree, very large even for a sequoia, still displays its maze of embedded fire scars, a warren of black chambers that entertained adventurous children in those days when parents were much more likely to allow their offspring to run free in the woods. I look for another old forest friend, a tree that stood adjacent to the front doors of the lodge's dining room, but it now lies shattered on the forest floor, finally brought down by the deep fire wounds that even half a century ago made us wonder about its ultimate stability. This is the tree that should have destroyed the old lodge dining room, parts of which dated back to 1915, but it did not. By the time the tree fell, the building was gone—razed and removed.

I scan the forest. Where fifty years ago I worked and lived in a busy village with a hundred cabins for rent each summer night and numerous other buildings housing the concessioner's operations and employees, I now see only quiet woods. The Giant Forest Lodge no longer exists. The removal and restoration is so complete that it takes considerable faith to believe that so much human infrastructure ever filled this forest. How this place came to be, and why it eventually could not be sustained, is a story worth remembering.

By 1920, the Big Trees were positioned to jump into the popular consciousness in a fundamentally new way. Three national parks now featured the sequoias, and although all three had existed since the late nineteenth century, the circumstances that defined them were undergoing profound changes. One change was the growing energy within the National Park Service, the federal agency founded in 1916 to oversee the parks. Stephen Mather, the new bureau's director, plainly saw publicity for the parks and their features as his number-one priority; the long-term political health of the parks, he believed, depended upon the creation of a substantial population of park users. At first Mather was largely stymied in his efforts to increase use of the park system by the entry of the United States into the First World War, an abrupt transition that occurred just months after the formation of the agency. Not until 1920, as it turned out, would the federal government be able to focus clearly again on nonmilitary domestic affairs. A second factor played an even larger role in changing the public's relationship with the trees: the arrival of the automobile age.

The mass marketing of inexpensive motorcars to the public had begun in 1908 with the introduction by Henry Ford of the Model T and the subsequent assembly-line concept that allowed the car to be produced cheaply. By the early 1920s, one could purchase a Ford automobile for less than $300. Running parallel with the advent of inexpensive automobiles was the good-roads movement—an effort that started in the 1880s as a campaign to provide better pathways for bicycles and then grew ever more important as the auto age took off after the turn of the century. Six weeks before President Woodrow Wilson signed the act that created the National Park Service, he signed the Federal Aid Road Act of 1916, the first legislation to designate federal funds for highway improvement. Quite literally, the National Park Service and the federal highway funding system came into the world as linked expressions of a moment in history. As the 1920s began, these two big changes in the national park world—the activation of aggressive federal marketing to increase park use and the advent of cheap automobiles and improved highways—made themselves felt in the three Sierra Nevada national parks.

The National Park Service's director, Stephen Mather, who had made himself rich in earlier years by marketing borax cleaning products to housewives, possessed a keen ability to anticipate public taste and develop marketable products. Now, even if he was selling national park experiences instead of cleaning powder, the need to find distinctive, attractive products had not changed. Mather understood intuitively that the swelling flood of automobile tourists would desire fundamentally different types of national park accommodations than had been developed in years past for well-to-do, railroad-based visitors. Automobile tourists would possess the means to be far more wide-ranging than their stagecoach-touring predecessors, and thus, Mather suspected, more sensitive to location and immediate experience as they sought accommodations. In this, Mather saw both a challenge and an opportunity. To merge these two factors successfully, the director grasped, accommodations must be placed in the most immediately scenic locales within the national parks. Inevitably this included the giant sequoia groves.

Mather had been thinking about such things since his first days with the Interior Department back in 1915. Some of his earliest goals for improving national parks accommodations focused on Yosemite, where a number of companies with mediocre facilities competed aggressively.

Mather envisioned, even at that date, that the best approach would be to consolidate all the small, independent operators into a single, better-funded company. With this goal in mind he brought in yet another concessioner, D. J. Desmond, with the goal that ultimately Desmond would become the park's primary concessioner. Desmond turned out to be a troubled operator, but over the next few years, before the money ran out, his company made a number of significant investments in Yosemite, building a modern hotel at Glacier Point and beginning development of the Yosemite Lodge near the base of Yosemite Falls.[1] By late 1917, the faltering Desmond Company had morphed into a successor corporation, the Yosemite National Park Company, and it was this iteration that opened the Big Trees Lodge in the middle of the Mariposa Grove in the summer of 1920.[2]

For several years, the Park Service had allowed what it called a "tourist camp" to operate in the lower edge of the grove, but this appears to have been a campground. Now, pursuing Mather's goal of catering to automobile-based visitors, it had more ambitious plans. The issue was not simply that more accommodations were needed in the Wawona area. In fact, just a year earlier, in 1918, the Washburn family of the Wawona Hotel had made major improvements in the form of a new two-story annex with 39 additional guest rooms. This brought the total number of available rooms at the hotel up to 126, all located within a half dozen easy road miles of the Big Trees. But Mather had other things in mind; he wanted to allow visitors to stay right in the midst of the best of national park resources, a strategy that he knew would prove popular. This had been the thought behind the new lodge in the Mariposa Grove. At the same time, he also approved similar new facilities at Tuolumne Meadows and at three locations in the backcountry of Yosemite.[3] And the new Big Trees facility had another benefit: it kept the visitors within the park during their overnight stay. The Wawona Hotel complex stood on private land that was outside park boundaries; Mather wanted to control the visitor experience, and that entailed keeping the visitors inside the park both day and night. That such a move might have a negative impact on the Big Trees does not seem to have even been considered.

The 1920 Big Trees Lodge in the Mariposa Grove took form in the very heart of the best part of the grove. It stood, in fact, just across the old touring loop road from the 1893 log cabin that had replaced Galen Clark's original "hospice." Built more like a modern High Sierra Camp than a formal lodge, the facility consisted of fifteen two-room tent cabins (grandly called "redwood cabins") and an open-air dining pavilion. This pavilion, framed in by a latticework of pole-sized timbers, quite literally circled the Montana Tree, a medium-sized giant sequoia. The tree rose column-like out of the center of the pavilion's roof. An adjoining tent served as the registration office.

Until the construction of the Big Trees Lodge, the Mariposa Grove had been spared for half a century from internal tourist development beyond roads and trails. Neither the state commissioners who had overseen the area from 1864 until 1906 nor the pre–National Park Service managers under the US Department of the Interior had ever seen fit to allow a lodge to be erected in the grove. Now, however, the grove was under new management pursuing distinctly different goals. And Yosemite was not the only target.

To the south of Yosemite, in Sequoia National Park, Mather followed a similar tack. A small and simple collection of tent cabins had grown almost organically in the heart of the Giant Forest after the first road opened to that site in 1903, and a dozen years later, the Interior Department had authorized the construction of a "hotel," but this really was simply a dining room to supplement the scattering of tent cabins. By 1920, Mather saw the opportunity to do more. The widow who had been operating the facility gave up her permit, and Mather brought in a new company that included some of the same partners who were involved in the Yosemite National Park Company. The new concessioner at Sequoia, operating as the Kings River Parks Company, immediately initiated a construction program, and from the beginning, there was no doubt that this development would occur in the heart of the grove. Mather wanted it that way, and the new concessions company had no reason to object. Over the next three summers, a complex called the Giant Forest Lodge began to develop on the site, and by 1925 the lodge had grown to have a visitor capacity of three hundred persons, all of them housed in either wooden cabins or tents tucked in closely among some of the park's most impressive giant sequoias.[4] This facility would continue to grow aggressively into the 1930s, protected and encouraged by Stephen Mather and his successor, Horace Albright. This

One of the first management actions taken in the Mariposa Grove by the newly organized National Park Service was to allow a lodge to be constructed in the heart of the grove. Note that this structure *surrounds* a giant sequoia tree. COURTESY OF YOSEMITE NPS LIBRARY

growth, however, would not occur without resistance from those few who were paying attention to the trees and their well-being.

In mid-May 1926, the Giant Forest Lodge hosted a distinguished scientific visitor. Emilio Meinecke, a forest pathologist, had come to Sequoia National Park at the express invitation of the National Park Service. Born in California in 1869 to German diplomat parents, Meinecke had been educated in Europe, receiving his doctorate in Heidelberg in 1893. His studies focused on root growth in plants. After a spell in Argentina, he returned to the land of his birth in 1909 and the following year accepted a position with the Bureau of Plant Industry, an agency within the US Department of Agriculture. Soon he was also serving as a consulting plant pathologist for another agency within that same department, the United States Forest Service. He would remain involved in forest questions for the remainder of his life. By the mid-1920s, Dr. Meinecke was known as California's leading plant pathologist, an expert on the health of both individual trees and forests. It was in this capacity that he arrived in Sequoia National Park on May 13, 1926.

Officially, National Park Service director Stephen Mather had requested Meinecke's visit to Sequoia National Park, but in reality the trip almost surely had been initiated by John R. White, a remarkable retired military officer serving since 1920 as superintendent of Sequoia and General Grant National Parks. Born in England in 1879, White had wandered the world while young. His experiences included serving in the Greek foreign legion, seeking gold and fishing for salmon in Alaska, and eventually joining the US Army as it fought to suppress an anti-imperialist revolt in the Philippines in the aftermath of the Spanish-American War. Transferring to the Philippine Constabulary, he rose to the rank of colonel, found himself invalided out with malaria, moved to California, and established residence there. He rejoined the US Army during the First World War, serving on General John J. Pershing's staff in France, then retired again and went looking for a new career. He chose the National Park Service. Director Mather initially assigned him to a ranger job at the Grand Canyon, but, recognizing his administrative experience, quickly transferred him to Sequoia and General Grant National Parks, where he became superintendent in July 1920.[5]

Mather hired White with the clear assumption that the naturalized Englishman would carry out the director's vision of developing Sequoia National Park with facilities for the rapidly increasing numbers of automobile visitors. For the next half-dozen years this is what White did. As this work progressed, however, White began to have doubts. With a new highway cutting its way through the Giant Forest, and both a growing lodge and automobile camping areas demanding space within the heart of the grove, White began to worry that all this development might well be damaging the very forest the visitors had come to see. Out of this concern came the invitation to Dr. Meinecke.

Meinecke spent only three days in the Giant Forest in 1926, but when he returned to his office in the Bay Area and sat down to document what he had seen, he nevertheless had a good deal to say. Meinecke had developed his reputation as a forest pathologist by emphasizing that healthy forests were best sustained by protecting the health of individual trees, and now he applied that logic to the sequoias of the Giant Forest. Noting that the trees were already effectively protected from fire damage and logging, he quickly moved to the issues of possible visitor impacts and root health. He found much of concern.

Meinecke's report on his visit to Sequoia National Park ran for nineteen pages. It represents, in many ways, the first attempt to document the impacts of tourism upon the Big Trees in a systematic way. He began by outlining both the overall goals of national park management and the fact that much harm had already occurred.

> If the main object involved in the setting aside of National Parks is the preservation, in their natural state, of selected regions of outstanding beauty or interest, then it must be recognized that the sins of past generations have left their indelible marks and have already modified natural conditions to a greater or lesser extent. This is particularly true in Parks of the type of Sequoia where the principal objects of interest and beauty are the Big Trees and the forest vegetation associated with them.[6]

Meinecke then moved on to talk about the inevitable conflicts that must arise from the challenge of making the trees accessible to "an ever growing enthusiastic but unconsciously predatory population of visitors." Completing his introductory remarks, Meinecke suggested what we would now term an interdisciplinary approach:

> The solution of the problem rests upon an understanding of the physiology and condition of the trees as individuals and members of a forest community and upon an analysis of tourist psychology and behavior.[7]

Getting down to the heart of his concerns, Meinecke then analyzed the trees as functioning organisms and commented upon how visitor use might damage them. Since the trees were already protected from fire by Park Service management, he believed that the greatest potential risks were associated with damage to the trees' roots systems. He described their "enormous water requirements" and their shallow and thus easy-to-damage root systems. Here, quite literally, was the root of the problem. Many aspects of park development damaged the fragile root systems of the Big Trees. Road construction severed roots of all sizes, as did the installation of underground utility systems. Visitors trampled understory vegetation, caused erosion that exposed tree roots, and compressed soils. All of this, the forest pathologist believed, weakened specimen trees.[8]

Meinecke's recommendations addressed both the forest as a whole and the management of individual trees. A critical starting point was to do no more damage. The Park Service must manage the Giant Forest so that further disturbances "are to be kept down to the lowest point compatible with the comfort and safety of visitors."[9] Individual trees needed to be protected from visitors by the judicious application of fencing and trail construction. Visitors would respond best, Meinecke believed, not just to regulations but rather to landscape design that met their basic demands to get close to the trees while at the same time protecting the trees' more fragile parts. The answer he proposed was to allow visitors to get close to the trees' dead fire scars (presumably not susceptible to additional damage) while protecting the more fragile parts of the trees where live bark connected with live roots. What this meant was not just fencing trees but also channeling visitors to approach them in the least damaging ways. Also important, he argued, was to continue the effort to move automobile camping areas, with all their unintended impacts, away from areas of high Big Tree concentrations, an effort that had already been initiated by Superintendent White.

Meinecke's Sequoia National Park report attracted the attention of Superintendent Charles G. Thompson in Yosemite National Park, and the following summer Meinecke visited the Mariposa Grove to study conditions there. In a report to Director Mather in September 1927, the forest pathologist documented problems and made additional recommendations that echoed what he written about Sequoia National Park the previous year.[10] In the Mariposa Grove, Meinecke was particularly concerned about the likelihood of root damage to some of the grove's best-known specimen trees as a result of soil compaction and erosion. He singled out the Grizzly Giant, the Texas Tree, and the Sheridan Tree. According to the forest pathologist:

> These trees, though they still continue to function, are obviously under a serious handicap from which there is no recovery so long as tourist

As shown in this early-1950s image from Sequoia National Park, visitors loved camping among the Big Trees. Few suspected how much such activity damaged the forests. COURTESY OF NPS, SEQUOIA AND KINGS CANYON MUSEUM COLLECTION

A 1930 effort to protect the Grizzly Giant Tree in Yosemite National Park surrounded the base of the tree with a network of metal stakes laced with barbed wire. To mask the presence of the wire, rangers planted shrubs. COURTESY OF YOSEMITE NPS LIBRARY

travel and camping continues to concentrate over the root spread of the trees.[11]

Meinecke's comments about conditions in the Mariposa Grove motivated Thompson to commence several efforts to protect individual trees. Perhaps the most drastic was a plan carried out in 1930 to ring the Grizzly Giant Tree with an ankle-high net of barbed wire attached to two hundred iron stakes pounded into the ground. Hoping to soften the resulting image, Thompson also ordered the planting of low bushes that he hoped would grow to hide the wires installed to protect the tree from visitors who loved to touch it.[12]

Pursuing another idea, Thompson also sent Alfred J. Bellue, the park's acting assistant forester, into the grove with instructions to study conditions there and make recommendations. In response, the forester found thickets of young trees, including sequoias, that needed to be thinned for optimum growth. He also was the first to note that the shade-tolerant white fir trees seemed to be growing so successfully that they were out-competing young sequoias.[13]

Bellue's observations ignited a debate within the Park Service about just what it was that the agency was trying to accomplish in its groves of Big Trees. Director Albright, trained as an attorney and certainly not an ecologist, opened the debate by reminding his park managers that the Park Service best managed nature by not managing it at all—that is, by a policy of noninterference. Thompson nevertheless pressed on with a program of thinning out the young trees within the Mariposa Grove while at the same time opening up vistas to show off the Big Trees. Sequoia National Park's superintendent, John R. White, dispatched Lawrence Cook, the park's chief ranger and a trained forester, to Yosemite to take a look at what Thompson was doing there. Cook reported positively to White. Over the following months, the conversation widened to include high-level officials across the agency. A basic question had been asked: Should the Park Service manage its special forests to achieve specific goals, or should it simply maintain a hands-off position? In all this, Yosemite's Superintendent Thompson continued to take a bold, even radical position:

In the Service there had too long persisted the ideal conception that in all the parks we are dealing with balances of nature, primeval conditions, and wilderness. In Yosemite, and in other parks excepting possibly McKinley, these words and phrases are mere shibboleths. For at least fifty years primeval conditions have been gone from every square mile of Yosemite.[14]

George M. Wright, the head of the Park Service's nascent research program, had followed these developments and now suggested that the agency literally put the idea of management of natural resources in sequoia groves to a formal internal trial, with a staff jury charged to resolve the question. Nothing came of that, however, and after director A. B. Cammerer (Albright had by now left the NPS to take a position in the corporate world) wrote to all involved that "management" of natural resources could not help but lead to "impairment," the debating parties moved on to other issues and left the question unresolved.[15] There it would remain for the next thirty years.

Arguments over philosophy and long-term goals were interesting to all, but in the meantime, practical on-the-ground issues still required answers. In both his Sequoia and Yosemite reports, Meinecke had discussed the impacts on the Big Trees of both road construction and campgrounds, but in neither case did he specifically mention the lodges that had been developed by the Park Service within the groves. This omission was not accidental; Director Mather had not wanted to hear about such things. His commitment to having overnight accommodations within the groves remained unshakable. This attitude, however, did not extend to Superintendent White at Sequoia. Even if Meinecke's analysis of visitor impacts to the sequoias had focused on campgrounds and roads, White could read between the lines; if campgrounds had the potential to do grievous harm to the Big Trees, why were lodges any less of an intrusion?

White spent the winter of 1926/27 pondering Meinecke's recommendations, and at the beginning of the 1927 visitor season he addressed his boss with ideas that must have seemed heretical at the time. The National Park Service, the Sequoia superintendent told Mather, must change its management directions within the Giant Forest. Limits must be placed

The Giant Forest Lodge of Sequoia National Park, repeatedly enlarged with the support of Park Service director Stephen Mather, became a prolonged source of controversy because of its location.
COURTESY OF NPS, SEQUOIA AND KINGS CANYON MUSEUM COLLECTION

on existing development at the Giant Forest Lodge and at the newly established housekeeping cabin complex called Camp Kaweah. Ultimately, White argued, a plan should be developed to remove the lodge cabins altogether from the heart of the grove adjacent to Round Meadow. A meeting followed between Mather, White, and Howard Hays, a primary investor in the park's concessions company, and White found his recommendations firmly rejected. Mather confirmed to Hays that commercial development in the grove needed to continue.[16] Rebuffed but not defeated (at least in his own mind), White had no choice but to retreat and bide his time.

John R. White, by this time a veteran national park manager, understood all too well the politics of his agency, and he settled down to wait for another opportunity to raise his concerns. In late 1928, Director Mather suffered a severe stroke, and he resigned the directorship in January 1929, identifying Horace Albright, his longtime number-one assistant, as his successor. With Mather gone, White resumed his internal opposition to the continuing development occurring within the Giant Forest. That summer he achieved his first success, convincing Albright to not allow Howard Hays and his company to build a new housekeeping cabin complex among the Big Trees at Round Meadow. Instead, White diverted the new cabins to a site without sequoias half a mile distant.

In 1931, White accelerated his campaign. When Hays came to him with proposals to build yet more cabins and a new dining room at the Giant Forest Lodge, White rejected the proposal and told the concessioner that he should begin making plans to remove his developments from the grove. Hays, long encouraged by Mather to keep building, erupted. Arguing that visitors came to the park to stay *beneath* the Big Trees and not just near them, Hays made clear his belief that his company would not be viable if forced to relocate its facilities. Director Albright, who had been an enthusiastic partner in Mather's development program, basically agreed with his concessioner but nevertheless worked to find a lawyer-like compromise. The answer, negotiated in late 1931, was to place a development limit of "200 pillows" on the Giant Forest Lodge complex. Albright also made clear, however, that lodging would remain in the grove and that White would have to accept that.[17]

Meanwhile, in Yosemite, Superintendent Thompson wrestled with similar questions. The Mariposa Grove, unlike Sequoia National Park's Giant Forest, did not represent the park's premier feature, but it nevertheless was receiving heavy visitor use and showing, according to Meinecke, evidence of resource damage. But Thompson was no John White.

Overseeing the grove in a way that undoubtedly made Director Albright comfortable, Thompson approved a new development program for the grove. It began with the construction of a modern water and sewer system in the grove, which involved the placement of over three miles of underground plumbing within the Big Tree area. He also authorized the replacement of the historic cabin in the heart of the grove. The only recognition of the grove's fragility was a decision to remove and relocate a campground from the lowest portion of the grove.[18]

Following closely on this work, the very snowy winter of 1931/32 brought a new issue to the fore. Heavy snow that season badly damaged the lightweight tent-frame structures at the 1919 Big Trees Lodge. Reflecting

Despite scientific evidence that tourist development was damaging the Big Trees of the Mariposa Grove, the Park Service allowed the construction of a new and more elaborate lodge in the grove in the early 1930s. COURTESY OF YOSEMITE NPS LIBRARY

the agency's philosophies, Thompson responded by authorizing a completely new Big Trees Lodge. To be located about a quarter mile from the previous site, but still within the grove and within easy strolling distance of the just-reconstructed cabin that marked the site of Clark's original "hospice," the new lodge was not a collection of tents with an open-air dining pavilion but rather a major self-contained building with twelve modern guest rooms and a fully enclosed dining room.[19] Arranged in an open V shape, the one-story lodge ran to a length of nearly two hundred feet and included a gift shop, nine employee dormitory rooms, and even a photography darkroom. If he were paying attention from afar, Meinecke must have shuddered at the notion.

The Park Service was pleased enough with the new Big Trees Lodge — designed by concessioner architect Eldridge Spencer — to place its plans in a volume issued a few years later to guide park-planners nationwide.[20] Still smarting in 1934 from the limits placed on expansion at the Giant Forest Lodge, Sequoia National Park concessioner Howard Hays did not mince words when it came to describing how unfairly he was being treated as compared to the concessioner in neighboring Yosemite National Park. In a letter to Director Cammerer, he complained:

> In Yosemite the Superintendent allowed the operator to build a completely new plant in the very heart of the Big Tree Grove....I was informed, moreover, that no limit had been officially placed on the future size of this place. It seems inconceivable that while [the] Superintendent in one park would be trying to put the brakes on development under Big Trees, the Superintendent in another park would be giving an operator free reign for development under the Big Trees.[21]

The Mather/Albright commitment to visitor services in even the most fragile of locations would not die easily.

Reflecting back on this period, it is hard not to see the huge contradictions embedded in the National Park Service's management of its sequoia groves. Despite clear evidence from scientific experts, and over the objections—at least in Sequoia—of park staff, the Park Service sustained its commitment to ensuring that visitor services ranked at the top of its list of priorities, continuing to assume that the natural features on these sites would survive without suffering significant damage. That it was already apparent that this was not true was simply ignored by the agency's top-ranking managers.

This era saw the first scientific analysis of visitor impacts on the Big Trees, and from this effort came the first clear evidence, provided by a scientific and unbiased observer, that preserving the groves and their specimen giant sequoias would entail considerably more management than the Park Service had previously understood. Tourism had the potential to damage the groves in ways that science could now document, and this led to a spirited exploration of just what the agency goals were for Big Tree management. Yet even with this information in hand, the Park Service continued to work aggressively to enhance and enlarge visitor development within the groves, and it continued in that direction even in the face of internal dissent from its own managers.

In the collision of science and politics, the Park Service ultimately selected the values that expressed what it thought most essential at the time.

It allowed Meinecke to reveal damage to sequoia roots from visitor traffic and road construction while it continued to maintain and even extend roads within both the Mariposa Grove and the Giant Forest. Campers caused trampling and erosion, the agency admitted, and the Park Service had no trouble relocating campgrounds to less sensitive sites, but lodges, on the other hand, were simply too critical to the public relations efforts to consider removing or relocating. The visitors who used them came from the better-moneyed classes, which Mather and Albright thought essential to the political survival of the parks. Meinecke knew this and stayed away from the question, at least in the explicit sense. Superintendent John White understood this too but plunged in anyway, only to be defeated repeatedly by the very concessions company that existed, at least in theory, to assist him in meeting the needs of the park visitors.

Through Emilio Meinecke, the National Park Service had made a small start in looking at the ecological question of what it would take to preserve the Big Trees for the long run, but the agency was not yet ready to hear the answer.

CHAPTER THIRTEEN
Words as Grand as Trees

There ought to be ghosts here, but on this sunny summer morning I can't find any. Perhaps I've come at the wrong time of the day. For decades, national park rangers wearing four-button-style wool military jackets in forest green and flat-brimmed Stetson hats stood here beside the campfire on summer evenings and told Big Tree stories. Now, decades later, all is quiet. The ranger-naturalists, the amphitheater that once stood here, and even the lodge that supplied the audience are all long gone.

This must have been a wonderful place to give campfire talks. Despite the intrusion of young trees onto the site, it is still possible to discern the physical outline of the outdoor amphitheater that occupied this site for nearly fifty years. A broad arc of wooden benches faced a raised earthen stage held in place by a fallen sequoia log. On the low hill behind the stage, a massive giant sequoia still rises more than 250 feet into the air. An old and deeply incised fire scar cleaves the base of the tree into two huge buttresses. The sequoia dominates the site in a way words cannot really capture.

But what of the stories that once filled the air here? What did Sequoia National Park rangers talk about years ago in the heart of the Giant Forest? That's harder to recapture than the physical contours of the amphitheater. No one recorded campfire talks in the 1920s and 1930s, and no scripts or even

For decades, specially trained "ranger-naturalists" helped national park visitors appreciate park wonders. One teaching aid was this giant sequoia cross section in front of the Yosemite Museum.
COURTESY OF YOSEMITE NPS LIBRARY

outlines remain in the park's archives. What does endure—our last, best window into that period—is the popular literature: the books and pamphlets that captured the perspectives of the era. To hear the ranger voices of campfires long past, we must turn to the books the rangers read.

In the 1920s, as automobiles flowed into the parks in ever greater numbers, the Park Service responded not just with programs of road improvement and campground and hotel construction but also with an educational initiative. All these new visitors, Director Mather believed, needed to become active advocates for the parks. This required that they be exposed to thoughts about what the parks contained and why they were so special. Out of this need grew a multifaceted educational campaign that included not only campfire programs presented by park rangers but also a new generation of literature about the Big Trees.

The opening contribution of the federal government to this literature was quite small. Beginning as early as 1915, the Interior Department began issuing annual visitor booklets for the national parks of the Sierra Nevada—free handouts that at first offered little about the Big Trees themselves.

The 1915 government booklet for Sequoia and General Grant National Parks—a publication that ran to forty pages—dedicated exactly two of those pages to the giant sequoias, and that short contribution was dominated by a table listing the sequoia groves to be found within the parks.[1] By the early 1920s, with the Park Service now staffed and in charge of the Sierra parks, the quality of the booklets improved, and so did their coverage of the sequoias. The 1921 edition for Sequoia and General Grant, now sporting a slick paper cover with a photograph of the Big Trees, included half a dozen pages of sequoia material, at the heart of which was a long quote from John Muir taken from his 1901 book, *Our National Parks*.[2]

Still, officials in the three parks had relatively little to say to help visitors understand the Big Trees under their care. The agency did, however, offer a five-cent supplementary booklet that provided additional information to interested visitors. Titled *The Secret of the Big Trees*, the twenty-four-page publication reprinted verbatim an article Ellsworth Huntington had written in 1912 for *Harper's Magazine*.[3] The Interior Department first published the booklet in 1913, and the Park Service continued to keep it in print until at least 1928. Reflecting Huntington's fieldwork during the summers of 1911 and 1912, the article focused on the age of the Big

By the 1920s, automobile tourism had become *the* way to visit the Big Trees of the Sierra Nevada. COURTESY OF YOSEMITE NPS LIBRARY

Trees and Huntington's conclusions about what could be learned from them about the role of the region's climate history. It was a good resource, but more were needed.

Other publications soon began to fill this gap. In 1920, a twenty-six-year-old Yosemite ranger named Ansel Hall unleashed the first of what would be a small personal barrage of publications about the national parks of the Sierra Nevada. A Californian by birth, Hall had taken a degree in forestry from the University of California, then joined the National Park Service as a ranger at Sequoia National Park in 1917. After serving in France during the latter days of the First World War, Hall returned to ranger duty in 1920, this time at Yosemite. Director Mather, who in those simpler times actually knew many of his agency's field staff, soon recognized in Hall a talent that the director could put to work in pursuit of both his educational and political goals. As a result, Hall became Yosemite's first federal "park naturalist." The young ranger soon turned out his first product: the *Guide to Yosemite: A Handbook of the Trails and Roads of Yosemite Valley and the Adjacent Region*.[4] This volume focused on Yosemite Valley, but within a year Hall had broadened his scope and begun to address the Big Trees of both Yosemite and Sequoia National Parks.

In 1921, Hall sent two books into the world, both aimed directly at national parks visitors. The first, based largely on information he had gathered during his 1917 season in Sequoia National Park, bore the lengthy title *Guide to the Giant Forest, Sequoia National Park: A Handbook of the Northern Section of Sequoia National Park and the Adjacent Sierra Nevada*.[5] The second, a more ambitious volume in many ways, came out as *Handbook of Yosemite National Park: A Compendium of Articles on the Yosemite Region by the Leading Scientific Authorities*.[6] Both addressed the Big Trees.

Hall wrote the Sequoia volume as a pocket guide to the most visited parts of that park and published it himself. The small book's chapter on the sequoias runs to only a few short pages, and the naturalist broke no new ground here, instead carefully referring the reader to other sources for more information. Muir remained the reigning descriptor of the Big Trees and, as Hall made clear, park visitors could find the appropriate portions of his work quoted at length in the free park brochure.[7] But by now Muir's sequoia writings had acquired a certain hoary age, having been published some thirty years ago and been based on fieldwork done almost half a century earlier. New material was available, including from scholars such as Huntington and Douglass, and to get at this information, Hall referred his readers to his other 1921 publication, the Yosemite *Handbook*.

Unlike his Sequoia National Park guidebook, Hall's *Handbook of Yosemite National Park* had not a single author but instead more than a dozen carefully selected contributors. The book's table of contents reads like a veritable who's who for the time and place. Contributors included eminent anthropologist A. L. Kroeber on the region's native peoples; University of California geologist A. C. Lawson on the origins of the physical landscape; and Joseph Grinnell (another UC contributor) on Yosemite's life zones, birds, mammals, and reptiles. NPS director Stephen Mather contributed a chapter on national park policy. Providing most of the material on the park's plants, including a chapter on the Big Trees, was Willis Linn Jepson, a professor of botany at the University of California.[8]

An accurate and cautious scholar, Jepson wrote an account that is as interesting for what it does not include as for what it shares. Jepson opened by describing the sequoias as "one of the most charmingly attractive features of Yosemite National Park," and after documenting the size of some of the best known of the trees, he moved into more scientific territory. Discussing their age, he incorporated what had been discovered by Huntington and Douglass, including the myth-busting facts that only a few trees reached 3,000 years of age and that no recorded tree-ring count went beyond 3,148 years. He summarized what was known of the geologic history of the species and explained the trees' ability to survive fires through the agency of their thick and fire-resistant bark. Reflecting the sensitivities of the wildfire-suppressing Park Service, however, Jepson made no mention of the tree's biological need for forest-thinning fires to reproduce.[9] Nor did he speculate about the uncertain origins of the name *Sequoia*.[10]

If Jepson, a university professor, focused on *Sequoia gigantea* from a primarily scientific point of view, others were more than happy to fill the public thirst for less weighty material. One of the first to address the need created by burgeoning national park tourism for a popular treatment of the giant sequoias was a young man named Rodney Sydes Ellsworth, a 1923 University of California graduate. Apparently as a seasonal job, Ellsworth had passed the summer of 1922 in the Mariposa Grove and become interested in the Big Trees. Back on campus, he began work on a manuscript that would tell the story of the trees in layman's terms, writing, as he put it, "a book by a tree-lover for tree-lovers."

Ellsworth's *The Giant Sequoia: An Account of the History and Characteristics of the Big Trees of California* came out from a minor California publisher in 1924.[11] An extensive appended bibliography suggests that the book likely began life as a senior thesis, but the final manuscript had nothing academic about it. The opening paragraph set a tone that the rest of the book would follow:

> The Sequoia is nature's most magnificent endowment. King of trees, it has no rival in size the world over, nor is it approached among living things of age. Noblest of all conifers it has the grandeur of granite and the solemnity of marble. Venerable in aspect, it savors of great antiquity....Is it a living survivor of an extinct age of monsters?[12]

In his rambling, prolix style, Ellsworth plowed familiar ground. After comparing the Big Trees to their coastal cousins, he moved into a prolonged discussion of the Mariposa Grove, describing both its famous features and its human history. His sequoias were long-lived not merely for biological reasons but rather because of their "unconquered will."[13] When it came to age, his bibliographic research captured the data produced by Huntington and Douglass, and he described estimates substantially in excess of 3,000 years as "assuredly absurd and fabulous."[14] Unlike even Jepson, the young writer also addressed that successful reproduction by sequoias required barren mineral soil and copious sunshine. Only in the closing chapter did Ellsworth drift into real trouble when he explored the origins of the tree's scientific name. He began well enough, telling the story of the tree's early nomenclature and how Endlicher successfully moved the species into the genus *Sequoia*, but then, after noting the lack of any firm information about where Endlicher got that name, he launched into the following:

Both Hooker and Englemann believed [the name *Sequoia*] derived from the Cherokee Indian, Sequoyah. At least it is edifying to know that Endlicher was an eminent linguist as well as a botanist. It is not improbable, then, that he was well acquainted with Sequoyah's colorful career and named the tree in honor of this aboriginal illiterate, this magnificent savage, who groped in darkness to give his people letters, and found the light.[15]

After that, Ellsworth provided a twenty-page-long biography of his Cherokee hero.

Despite this ill-informed closing, Ellsworth's 167-page book captured the essence of most of what was known about the Big Trees in the early 1920s and translated it into popular form. Automobile tourists passing time in Yosemite's Big Trees Lodge or the Giant Forest Lodge in Sequoia National Park now had something of substance to read about the sequoias besides John Muir. We don't know today how well Ellsworth's book sold, but his success may be gauged by the fact that enough copies have survived to make the volume relatively easy to obtain even today on the used-book market.

As the 1920s progressed, so did the certainty ascribed to the Sequoyah story. The proposed connection between the big red trees and a "noble red man" struck a powerful chord. With the Indian Wars now a full generation in the past and the continent's native inhabitants fully "neutralized," America had begun to romanticize the native peoples it had worked so hard in the nineteenth century to subjugate and displace. A narrative that somehow the sequoias of California had been named by an Austrian botanist in honor of the Cherokee Sequoyah—a proposition totally without supporting evidence beyond the fact that the two names were homonyms—thus conformed perfectly with America's new hunger for positive stories about the peoples it had almost destroyed.

Enter Herbert Earl Wilson. Born about 1892, Wilson had, by the early 1920s, found a way to make a living celebrating America's romantic infatuation with Native Americans. Often dressing as an Indian, Wilson attended public events and gave talks at national park lodges. In 1922, he published his first book, *The Lore and the Lure of the Yosemite*, which, despite its title, was dominated by a collection of Native American stories, most of them taken from Galen Clark's 1904 book on that subject. The other sections of the book offered little that was new; his short section on the Big Trees came straight out of Muir.[16]

By the middle of the 1920s, Wilson had shifted his attention to Sequoia National Park. Out of his experiences in the southern park came his second book, *The Lore and the Lure of Sequoia: The Sequoia Gigantea, Its History and Description*.[17] This time, Wilson managed to capture and perpetuate nearly all the classic attitudes and myths that had been accruing about the sequoias since the 1850s. The General Sherman Tree, for example, could be turned into lumber sufficient to build housing for "2,470 people."[18] The same famous tree was "well on its way to five thousand years of age."[19] In another section he speculated, without any evidence or apparent source, that "before the uplift of the Sierra Mountain, the Sequoia forest covered all the region which they now occupy."[20] As for fire, Wilson echoed the dominant views of his time, describing fire as a major threat, but also ascribing to the Big Trees an almost miraculous ability to survive these assaults. Like so many others over the years, Wilson found this deep-seated capacity to survive a source of inspiration:

> As we walk or rise through these beautiful Sequoia groves, let us give homage to these monumental, living gifts of nature, standing thus throughout the centuries, watching generations upon generations of fir and pine grow from infancy, attain old age, and then go in death to the earth that mothered them.[21]

Wilson, in his Sequoia Park volume, also erased all possible doubt from the Sequoyah story.

> Monuments of stone, bronze, and marble have been sculpted for men of great deeds, but none could be erected to this Cherokee Indian, Sequoia [sic], as great as this sublime living monument, the Sequoia Gigantea, which Nature has built and which has been named for him that he shall be known forever throughout the world.[22]

Herbert Earl Wilson may not have been a scholar, but he knew how to produce a book that spoke to the increasing number of tourists that automobiles brought to the giant sequoia groves of the Sierra Nevada. He told visitors what they wanted to hear—stories about noble Indians, ageless trees, and friendly wildlife. He confirmed their romantic need to be told that the sequoia trees bore a name that helped right some of the historical wrongs of earlier generations. In short, he captured and strengthened the popular myths of his time. And, by all evidence, the books sold well.[23]

By 1930, however, the public mood had again shifted. The calamitous arrival of the Great Depression knocked many of the excesses out of national park tourism and ushered in a more somber time. Reflecting this new outlook were two new books about the Big Trees.

George Stewart's *Big Trees of the Giant Forest* can only be described as a labor of love.[24] Forty years earlier, Stewart had led the campaign to create a giant sequoia national park in Tulare County, and he had never lost his affection for the Big Trees. Now, approaching the end of his life, the attorney and newspaper editor set out to record what he had learned about the trees. The book's subtitle laid out his primary goal: *Their Life Story from the Blossom Onward*.

Unlike many of his contemporaries, Stewart had the discipline to write only about things he knew by personal experience, and he focused the book entirely on a single grove—the Giant Forest of Sequoia National Park. No one since Muir had attempted to write about sequoias by following them through their life cycle while exploring the challenges they faced and how they overcame them to achieve such great age and size. Muir had grasped many of the essentials of this story during his relatively short time among the Big Trees, but Stewart was drawing upon decades of affectionate observation. His collective "biography" of the sequoias, for that is what it was, followed their life stories through all phases from germination to eventual death. Along the way, he referred always to actual trees growing in the Giant Forest.

Stewart provided no bibliographic references and seldom referred to other, earlier writers, but it was apparent that he had read the existing literature. In most ways, he followed the accepted interpretations of his day. Inevitably, as the book moved toward its conclusion, Stewart returned to the General Sherman Tree. He took a notably cautious approach, noting that the tree was not as old as many surmised, while admitting that he believed it to be "nearer four thousand years."[25] Although he must have been aware of the tree-ring work of the previous two decades, Stewart made no mention of it in his discussion of the age of the Big Trees.

In many ways, Stewart wrote an elegant and well-informed book about the sequoias, but circumstances severely limited its reach. Not only did the book come out in the first full year of the Great Depression, but it also found itself in direct competition with another book that came out at almost the same time: *Big Trees* by Walter Fry and John R. White, a volume that would come to be recognized as the first enduring sequoia classic since Muir.[26]

Fry and White made a strong writing team. Both had served terms as superintendent of Sequoia and General Grant National Parks, but otherwise their experiences brought very different skills to the project. Fry was a self-taught but consummate naturalist. Born in Illinois in 1859, he came west with his wife to settle in Tulare County in 1887. At first he worked for the Southern Pacific Railroad, but by 1895 he had established a cattle ranch in the Three Rivers area near the boundary of the new Sequoia National Park. Early in his California days he had signed on as a lumberman with the Smith Comstock logging show near the Grant Grove, but he quit after only a week, having spent his first day off counting the several thousand rings on a Big Tree that he had just felled. In 1901, he finally found his calling when he went to work for the national park, first as a road crew foreman and then as a civilian ranger. In 1914, after the US Army pulled out of Sequoia National Park for good, he became the park's first civilian superintendent, a position he would hold until 1920. During these years, his work kept him almost continuously outdoors in the Sierra. In his later years he would estimate that in fifteen years he rode on horseback more than 50,000 miles along park trails, and all that time he was conscientiously scribbling notes about natural events and patterns. On July 1, 1920, Fry stepped down as superintendent of the two southern Sierra national parks to be replaced by the man who would become his coauthor for the book *Big Trees*—John R. White. Fry's connection to the parks did not end, however, as he then assumed the position of park magistrate, an undemanding job that allowed him to pursue his main interests: studying and writing about nature.[27]

Together, John R. White (left) and Walter Fry wrote *Big Trees*, a 1930 volume that served for a generation as the best book about the giant sequoias. COURTESY OF NPS, SEQUOIA AND KINGS CANYON MUSEUM COLLECTION

In every way except for his love of the sequoias, John R. White was a profoundly different man than Walter Fry. As mentioned in the previous chapter, White's expertise initially fell mainly into the areas of organization and management, but he found the sequoias to his liking and soon put down deep roots. At the same time, Fry, now "Judge Fry," took White under his wing to teach him the natural history of the sequoias. By the late 1920s, White knew a good deal about the Big Trees. Certainly this education helped fuel his previously described doubts about the continuing development within the Giant Forest. White also had another skill; he knew how to write.

Big Trees came out from Stanford University Press in 1930. Handsomely produced, the 100-page volume reflected the strengths of its authors—Fry's intimate knowledge of the trees and White's consummate communication skills. The two wrote for a general readership but, reflecting the book's university press origins, the volume contained real substance.

From the opening of the first chapter, White set a tone that seemed to capture the essence of the giant sequoia experience.

> At the entrance to a grove of trees the Romans would place the inscription: *Numen inest*, "God is in this place." And when the traveler to Sequoia National Park turns his back upon roads and automobiles, and directs his feet up the trail that leads through the firs, pines, and sequoias to the Congress Group, it is almost inevitable that he pause for a moment when passing between the four trees that form The Cloister, feel there the spirit of the forest, mellowed by the centuries, and know, although there is no bronze or marble plaque to tell him so, that "God is in this place."[28]

But the true strength of the book came from Judge Fry's immense fund of firsthand observations. As the text moved beyond the opening historical chapters, Fry's decades of careful note-taking began to show. The naturalist documented, for example, that he had counted the growth rings on no less than 1,982 Big Trees and found only two older than 3,000 years and, in fact, only 227 older than 2,000 years.[29] Moving on, he shared that the oldest count he had personally made provided an age of 3,126 years, but since that tree was only about twenty-six and a half feet in diameter, it was logical to assume that the biggest trees could be older.[30]

Using his forty years of personal study as a foundation, Fry wrote at length about how the tree's growth rings reflected the vagaries of the Sierra's climate. The same long experience allowed him to write accurately about the pollination habits of the Big Trees, including their system of producing pollen and fertilizing their female cones during midwinter. One of Fry's most interesting experiments began in 1905, when he encased a dozen freshly matured sequoia cones in wire mesh and then tracked them for the next fourteen years as they continued to hang on the tree. Fry was apparently the first to document that sequoias, unlike most other conifer trees, did not fully open their cones upon maturity or immediately distribute seed. Instead, as Fry learned between 1905 and 1919, the cones remained on the tree for many seasons and dropped their seeds only slowly.[31] In short, here was a man who had paid close attention to what was going on around him.

In perhaps the strongest chapter in the book, Fry and White explored "The Death of a Giant." Here, Fry was able to bring into focus his decades of observing the effects on the trees of both lightning and fire. He described fire as not simply the tree's mortal enemy but instead as something much more complex. Different fires, Fry wrote, burned among the groves in differing ways and with differing results. He outlined the behavior of what he called ground fires, crown fires, and underground fires and how each affected the Big Trees. All the larger trees, Fry documented, bore evidence of past fire exposure, and he went so far as to propose that the fire frequency in the groves prior to the area's national park status had been about once every twenty-five years, an estimate amazingly close to the number fire-ecology researchers would validate more than fifty years later.* Fry, again using his decades of personal observation, documented how basal fire scars came into being and how basal fires damaged and sometimes killed the tops of Big Trees. He also noted that seedling sequoias responded well to the disturbed conditions created by fires.

In his tens of thousands of miles of trail patrol within Sequoia National Park, Fry had seen things that perhaps no one else had ever witnessed. Now he could share these moments. Describing a lightning storm he observed in October 1905, Fry wrote:

> Never shall I forget that sight. For when the lightning hit the tree we could see clear daylight through the opening of the cut, and broken chunks of tree scattering everywhere; and while the two split portions of the tree gaped wide apart, the cut-off top of the tree was for a moment poised erect in the air above. Only for a moment was this strange spectral tree top suspended in midair. Then it dropped straight downward between the two open slabs, which clamped tightly upon it. There it is held to this day....[32]

Working again from Fry's intimate knowledge of the groves and from information gathered by White about the neighboring national forests, the

---

* Today, the National Park Service estimates that the natural fire frequency in the groves prior to the arrival of Euro-Americans was about once every fifteen to twenty-five years.

two authors also offered in an appendix a new list of giant sequoia groves together with their relative sizes and significance. For decades most students of the Big Trees had described their range as consisting of roughly thirty groves. In 1923, Willis Jepson, California's leading botanist, offered the number of thirty-two in his popular book *The Trees of California*.[33] Fry and White took a fresh look at this question, and they began by doing something that no one had ever done before: they offered a definition of what constituted a "grove."

> We have carried as distinct groves those parts of the Big Tree forests which are clearly separated from other areas of Big Trees by a belt of forest at least half a mile wide in which no sequoias occur, or which are separated by some natural division, such as a rocky ridge, that clearly defines the forest area.[34]

Working from this definition, the two authors came up with an entirely new list of giant sequoia groves. By their count, the Sierra contained not thirty-two groves but actually seventy-one distinct giant sequoia stands, which varied in size from tiny groves with a handful of trees to large stands containing hundreds or thousands of large trees.[35] This list would become the starting point for all subsequent efforts to enumerate the giant sequoia groves of the Sierra Nevada.[36]

As a team, Walter Fry and John White wrote a book that took popular understanding of the giant sequoias to a new and unprecedented level. Fry's unmatched fund of firsthand knowledge together with his willingness to initiate and follow through with long-term natural history projects (many were not exactly hard science) had made the difference when combined with White's strong compositional skills. *Big Trees* rose quickly to become *the* mainstream book on the giant sequoias, a status it would hold for forty years as it went through a major revision in 1938 and at least seven printings.[37] More than any other twentieth-century book, *Big Trees* was the defining work for public understanding of the sequoias.

What does all this tell us? Returning to those long-ago campfire talks where we began this chapter, we can now imagine something of what was being shared during those mountain evenings. The early ranger-naturalists of the National Park Service must have found themselves pulled in contradictory directions. To hold the interest of their visitors, these uniformed men needed to tell good stories—stories that resonated with their audiences. The popular literature contained many such narratives, and national park visitors loved to hear about how the trees were essentially immortal, about how trees still standing had already been giants when Christ walked the earth, about how the trees bore the name of a noble Native American.

But the real story was both more complex and less satisfying, and the ranger-naturalists wrestled to present a more complete vision—the vision they found in serious books like Fry and White's *Big Trees*. Over time, a new orthodoxy evolved. The giant sequoias, the largest and oldest of all living things, were best understood as awe-inspiring organisms that transcended the limits found in all other vegetable forms. They possessed an indomitable will to endure and, with adequate protection, did have the potential to live almost indefinitely. Best seen as endangered relicts of another age, they grew in only a very few places and reproduced only with luck. From humans, they deserved both physical protection and profound respect as examples of the most amazing of God's handiwork. That much of this interpretation was anthropomorphic cultural projection made little difference. The mythology of the sequoias had achieved a degree of perfection that rivaled the trees themselves.

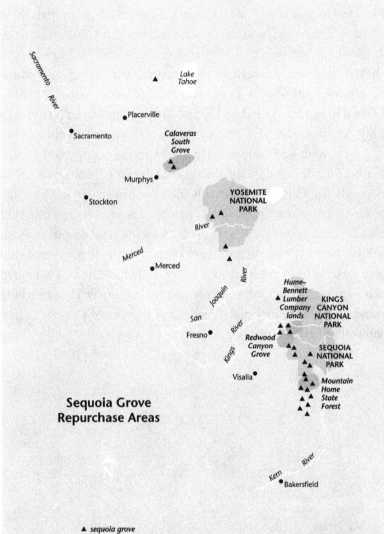

CHAPTER FOURTEEN
Belonging to All

The narrow dirt road—a minimal two-track—leads me deeper into the forest. I creep along in my 4x4, although the going is not all that rough. There's plenty to see. Around me an open and sunny forest displays not only copious pine and fir stands but also numerous monarch sequoias. Every few hundred yards I come to another unmarked junction. I have only a general sense of where my destination lies, so I proceed by trial and error.

A quick look out the window confirms that this is not a national park. If it were, I would not be exploring this grove in my 4x4. Occasional rotting piles of logging leftovers reinforce the same message. This forest has been gently logged, harvested in what looks to have been a sustainable manner. Not only do big sequoias remain here, but so do many other large trees of several species. Decades of carefully executed selective logging has left this forest largely intact and without the thickets of brush and younger trees that typify so much national forest land where logging was more aggressive. I pass a small pullout where others have parked, then come quickly to the end of the road—a turnaround where logging trucks loaded in years past. Taking another look at my maps, I reverse direction, creep back to the wide spot, and park. I guess I'm here. A faint unmarked trail—really no more than a trace—leads gently uphill through waist-high bracken and thimbleberry. I start walking and a Big Tree soon comes into view. A very big Big Tree.

A small wooden sign, almost hidden in the low, green thicket that surrounds the tree, confirms that I have found my quarry. The trunk of the Genesis Tree rises massively before me. My references tell me that this seldom-visited giant, growing quietly among its many sequoia siblings, is a member of a select and storied group. It is one of the ten largest sequoias, a list that includes such famous monarchs as the General Sherman and General Grant Trees. On this list the Genesis vies with several other lesser-known trees for the title of least-visited. As I have discovered today, just finding this tree, even though one can drive to within several hundred yards of it, requires navigation skills, perseverance, and luck. It goes without saying that no one else is here. The almost overgrown trail leading up from where I'd left my truck confirms that this must usually be the case.

I study my maps once again. The boundary of Sequoia National Park is a half-dozen miles to my north. I am much closer to the boundary of that other federal reservation, the Sequoia National Forest, but I am not within it either. This land—now a part of the Mountain Home State Forest—has a complex history. Sold before 1893 under the auspices of the Timber and Stone Act, this tract eventually came under the control of Thomas Hume, the same lumberman who controlled the Hume-Bennett Lumber Company that owned the Converse Grove as well as the numerous sequoia groves east of Hume Lake. How this grove avoided the fate of those forests and ended instead belonging to the people of California is the story we must now take up.

By the early 1920s, the giant sequoias of the Sierra Nevada had cemented their status as national icons—defining organisms recognized as the largest and oldest of the earth's many life-forms. The trees had now been famous for two-thirds of a century. Three national parks celebrated their uniqueness, as did a growing body of scientific and popular literature. The wave of logging that had gained so much momentum in the late nineteenth century and brought down so many Big Trees had wound down after proving to be unprofitable. Moreover, with California quickly moving away from its pioneer roots, the felling of the sequoias had become less and less acceptable to the urban populations that now comprised a steadily growing proportion of the Golden State's citizenry. The creation in the Sierra Nevada of national parks, forest reserves, and ultimately national forests had brought many sequoia groves under permanent management by the federal government, but many other trees, some of them well-known and loved, remained in private ownership as a result of nineteenth-century federal land laws. Increasingly, the fate of these private lands—and their trees—began to attract public attention.

One of the first public figures to address this question was none other than Stephen Mather, the first director of the National Park Service. Even as he began his work in the Interior Department in 1915, Mather was aware that the government, prior to the establishment of the parks, had sold some key Big Tree tracts within the Sierra Nevada national parks. The Giant Forest of Sequoia National Park was particularly at risk. Even though the Timber and Stone claim filed on the grove in 1885 by the Kaweah Colony had never been validated, hundreds of acres of prime giant sequoia forest had been sold under the provisions of another federal statute, the Swamp and Overflow Act. At the center of the problem were some 670 acres owned by two Tulare County ranchers who had grazed the land intensively but never allowed any logging. Mather put his considerable political talents to work and in 1916 achieved results in the form of Congress appropriating $50,000 for the acquisition of the lands and their trees. Unfortunately, the ranchers wanted $70,000. Believing that Congress was not likely to increase the appropriation, Mather set out to raise the balance privately.[1]

Gilbert Grosvenor, managing director of the National Geographic Society and editor of its magazine, occupied a spot very near the top of Mather's list of national park friends. From the beginning of his campaign to create a bureau of national parks, Mather had cultivated Grosvenor. The editor had been a key participant in Mather's 1915 High Sierra pack trip, and had responded by using his magazine to advance the park bureau campaign. In April 1916, as Mather moved that campaign into high gear, Grosvenor's magazine published "The Land of the Best," an article that featured the national parks and filled the entire issue. Serving as the frontispiece was a twenty-four-inch-tall foldout photograph of Sequoia National Park's General Sherman Tree. In the caption, Grosvenor pulled no punches:

> THE OLDEST LIVING THING: Towering a giant among giants, the oldest living thing that connects the present to the dim past, majestic in its mien, its dignity and its world-old experience, the "General Sherman Tree" is the patriarch of the Sequoia National Park of California. It was already 2,000 years old when Christ was born. In the age when the known world was rocking in the throes of the Trojan Wars and the time that history tells us marked the Exodus of the Hebrews from Egypt, this greatest of *sequoia gigantea* was a flourishing sapling some twenty or thirty feet in height, and truly under the special care of the Creator, Who held it safe from the lightnings of His wrath as He did from the attack of earthly enemies.*[2]

This enthusiasm came in part from Grosvenor's visit the previous summer to the Giant Forest and its largest tree. Now, with only limited time available to supplement the congressional appropriation intended to purchase the private lands within the heart of the Giant Forest, Mather turned again to Grosvenor and the National Geographic Society. He was not disappointed. In November 1916, the society donated the remaining $20,000 that brought some of the best of the Giant Forest back into public ownership and national park management.[3]

The 1916 acquisition and protection of the private Big Tree areas in the Giant Forest was only the beginning of what Mather had in mind for Sequoia National Park, however. After August 1916, with the National Park Service now in existence, Mather moved forward with political goals related to a number of existing federal parks. High on this list was enlarging Sequoia. As created in 1890, the park focused on the sequoia groves that were to be found in the mid-altitude region of the Kaweah River watershed, the very country John Muir had explored in his 1875 giant sequoia trip. Immediately to the east of the national park, however, could be found

---

* Today, the age of the General Sherman Tree is estimated by the National Park Service to be less than 2,500 years, rather than the 4,000-plus years suggested in this quote.

the highest and some of the most scenic mountains of the forty-eight contiguous states. Of equal interest to Mather was the great glacial canyon of the Kings River, which occupied the mountain country northeast of the existing park. Even by 1915, when he led his political pack trips into this region, Mather was describing this mountain wonderland as the "Greater Sequoia" region, a land that needed to be added to the existing park to make it whole.[4]

Theodore Roosevelt had died unexpectedly in January 1919, and that summer Mather was marketing the concept of a "Roosevelt-Sequoia National Park" that would be a memorial to the ex-president. As proposed, the new park would take in a huge swath of High Sierra, including the great canyons of the Kings River. The proposed boundary would also acquire the eastern sequoia groves controlled by George Hume's consistently unprofitable Sanger Lumber Company.[5] This is what brought Mather as well as Forest Service chief Henry Graves to Hume Lake in July 1919.[6]

George Hume, eager to sell his lands, met Mather and his party and organized and hosted a tour of the sequoia groves to the east of Hume Lake. The party traveled along the logging railroad in an open-air flatcar outfitted with benches. In the end, neither Mather nor Hume got what they wanted from this encounter. The Forest Service, determined to maintain agency control of the Kings River watershed, continued to oppose Mather's campaign to enlarge Sequoia National Park, and when a compromise was finally achieved in 1926, it did not include any Kings River lands. George Hume's sequoia trees remained in private ownership through the 1920s.[7]

Other efforts to bring sequoias back into public ownership did better. In the Tule River watershed, within the extensive Mountain Home Grove of giant sequoias, the San Joaquin Light and Power Company had come into the ownership of 160 acres, land that had been acquired to provide clear-grained Big Tree lumber for the flumes the power company intended to build. The flume project did not work out, however, and Natalie Harris Hammond, whose husband had been involved in the project as an engineer, argued that the two hundred sequoias on the tract were too beautiful to destroy. In 1923, Allan Balch (another power company official) and his wife, Janet, purchased the tract with the specific intent of turning it into a public park to be managed by the local county government. Tulare

County formally took control of "Balch Park" and its Big Trees in December 1930.⁸ It remains a county park today.

An event of greater significance was the long-delayed public acquisition of the original discovery grove of Big Trees at Calaveras. Lumberman Robert Whiteside had acquired the grove, together with much surrounding timberland, in 1900, but he did not allow logging in the grove. Instead, during the 1910s and 1920s, Calaveras remained just what it had been since the 1850s: a privately owned tourist destination. The old Sperry's Hotel continued to welcome guests, and the famous named trees still worked their magic on visitors. Several attempts to bring the grove into public ownership were no more successful than the national park effort of 1900. These abortive efforts included federal legislation in 1909 that authorized the Forest Service to acquire the land by trading acreage with Whiteside and thus create a Calaveras Bigtrees [sic] National Forest. Whiteside refused to cooperate. Even a visit by Stephen Mather in March 1924, after which he recommended that the grove be purchased by the federal government, led to nothing. Matters finally shifted in the summer of 1926, when it became public knowledge that Whiteside had entered into negotiations to sell 12,000 acres of his California holdings to the Pickering Lumber Company. These lands, located south of the Stanislaus River, included the undeveloped and seldom-visited South Calaveras Grove. Word also went around that Whiteside was in talks to sell the 1,760-acre parcel north of the river that contained the discovery ("North") grove. This was alarming news.

Rising quickly to the occasion, local banker and chamber of commerce president Désiré Fricot stepped up and personally obtained a ninety-day option on the North Grove from Whiteside. Fricot's goal was to find a way during that time to move the Big Trees toward public ownership. The bank president failed to raise the necessary funds in time, but his efforts generated considerable publicity and set others into motion. The first response to this flurry of concern was that Whiteside agreed not to sell the North Grove to any private party and instead hold it until it could be acquired by a public agency. At the same time, a handful of leading local citizens, including Fricot, formed the Calaveras Grove

Association, an organization intended to facilitate public acquisition of the Big Trees in both the North and South Groves. Finally, responding to the Calaveras situation as well as other opportunities, there began an effort to create a California state park system, with the Calaveras Big Trees as one of its founding units.

All of this looked promising, but then the Great Depression hit, and the association's fundraising program stalled. The association abandoned its effort to acquire the South Grove and focused all its remaining energy on the discovery grove, but success even there seemed in doubt. Eventually Whiteside set a deadline of March 15, 1931. If the State of California did not come through by that date with a firm offer, he would sell to the lumbermen.[9]

At the very last minute, funds for the purchase finally came from the Save the Redwoods League, which succeeded in obtaining financial support from John D. Rockefeller, Jr., who had begun to donate generously in these years to conservation projected all over the nation.[10] This offer was for matching funds, however, and it took a final personal loan of $7,000 to the association by board member J. C. Sperry (the son of the Sperry who had once owned the grove) to complete the purchase. The deal closed on June 24, 1931. The State of California now owned the North Grove of the Calaveras Big Trees, the place where our giant sequoia story had begun nearly eighty years earlier. Barely two weeks later, on July 5, a celebration marked the formal dedication of the area as a state park.[11] Public support for the giant sequoias had saved yet another grove.

The advent of the Great Depression, which so severely challenged the Calaveras Grove Association, changed the playing field when it came to the acquisition of privately held sequoia groves elsewhere. The almost complete collapse of the construction trade depressed the price of timberlands nationwide at the same time that it forced many lumber companies to shed assets. What resulted was an unprecedented opportunity for the government to purchase giant sequoia lands.

In Yosemite, the hard times of the early 1930s brought the company that owned the Wawona Hotel to its knees. The Washburn family had controlled the property since the 1870s and had overseen the construction of the Wawona Hotel in 1879. Now, several generations later, it appeared that the time had come to sell their more than 2,600 acres of private land, together with the hotel complex that stood upon it. A few years earlier, the Yosemite National Park Company—the concessioner that operated both the Big Trees Lodge in the nearby Mariposa Grove and the hotels in Yosemite Valley—had made an offer for the property, but the Washburns had not then been ready to part with their long-time home. Now, hard hit by the Depression, the park concessioner was no longer in a position to make an offer, but another entity stepped forward. The Wawona Basin, including the hotel, had never been a part of Yosemite National Park, even though park lands immediately adjoined the Washburn tract. The Park Service had long believed that it would make sense to extend the park to include the basin and its facilities, and the moment had come to make that move. In August 1932, the federal government purchased the 2,665 acres that contained the hotel and all the other private lands that Galen Clark and his successors had accumulated. The government got a bargain, paying only half of the $150,000 price for the key tract; the other half came from the park concessioner, who gained the right to operate the now government-owned hotel for the next twenty years.[12] On August 26, all the lands in question became a part of Yosemite National Park. The acquisition of Wawona added no native sequoias to the park, but it did bring under Park Service management the visitor infrastructure that had long supported the neighboring Mariposa Grove. Henceforth, Wawona and the Big Trees next door would function as the unitary destination most visitors had always perceived them to be.

Three years later, a much grander sequoia prize came into government hands. After the collapse of his effort to sell his timberlands to the federal government for inclusion in an enlarged Sequoia National Park, George Hume continued to seek a way to profitably divest his family of the more than thirty square miles of Sierra Nevada forest still under their control. In July 1926, the same month that Congress extended Sequoia National Park to include Mt. Whitney but not the Kings Canyon country, a forest fire consumed six miles of the lumber flume that connected Hume Lake with Sanger in the San Joaquin Valley, thus further isolating Hume's timber holdings from the commercial market. The following year, Hume found a buyer for the mill and buildings at Hume Lake, but he continued to be saddled with the extensive forestlands that had once been the prime asset of the Hume-Bennett Lumber Company.

Long conversations followed with the US Forest Service and eventually, with the Great Depression coloring the situation, the agency came to an agreement with Hume to purchase his approximately 20,000 acres for addition to the Sequoia National Forest. The deal, consummated on April 8, 1935, cost the government $319,276.75, or about $14.93 per acre.[13] Fifty years earlier, the government had sold most of this land for $2.50 per acre. Now, with much of the old-growth forest removed, the same lands came back into public ownership. Included were both the stump fields of Converse Basin and the half-dozen sequoia groves that had sent their logs to the Hume Lake mill before that effort collapsed. The largest private owner of Big Tree forest in the Sierra Nevada had finally found a way to liquidate its holdings.

Meanwhile, the campaign to create a national park featuring the great glacial canyon of the South Fork of the Kings River had come back to life. The political compromise of 1926 that enlarged Sequoia National Park to the east but not to the north had put the issue on hold for a decade, but in the mid-1930s the Sierra Club had revived the issue. Most of the argument was about the High Sierra, but publicity had also begun to circulate about the need to do something to protect the giant sequoias of Redwood Canyon and Redwood Mountain, a very large Big Tree grove south and east of General Grant National Park. Muir had explored this grove in 1875 and had found Hyde's Mill busy in the vicinity cutting giant sequoias. Later, after it was surveyed, the thirty-six-square-mile township that included the grove had been a primary target for Timber and Stone Act claimants. By 1900, when George Sudworth of the US Department of Agriculture surveyed private ownership of Big Tree lands, he reported that the township contained some fifty-nine separate private tracts.[14] In the early years of the twentieth century, these holdings slowly consolidated, yet, almost miraculously, no major logging occurred. Only a handful of Big Trees fell, mostly to be split into grape stakes and fence posts.

By the 1920s it had become apparent that the Redwood Mountain Grove, despite its complex ownership history, stood as one of the Sierra's premier giant sequoia forests. The Big Trees covered well over 4,000 acres, and the grove contained numerous specimen trees. Among these was the Hart Tree, raised as a contender in the 1920s for the title of largest tree in the world. In 1931, when an engineer named J. W. Jourdan made comparative measurements of several of the largest of the giant sequoias for the Fresno Junior Chamber of Commerce, the Hart Tree was one of the four measured. As might be surmised from its sponsorship, this effort had a quasi-political origin. Jourdan's four trees were what were generally believed to be the largest tree in Sequoia National Park (the General Sherman), the most massive in General Grant National Park (the General Grant Tree), the leading national forest contender (the Boole Tree in Converse Basin), and the Hart Tree in Redwood Canyon. Jourdan's measurements placed the trees in relative order—General Sherman first, followed by the General Grant Tree, the Boole Tree, and then the Hart Tree—and this list soon morphed inaccurately into "the four largest trees." The inclusion of the Hart Tree, which measured significantly smaller than the other three on this list, has been interpreted as an effort to draw attention to the superlative Redwood Mountain Grove by those who hoped to bring it back into public ownership.

Despite its size—the largest by acreage of all the groves except Converse Basin—the Redwood Mountain Grove had largely escaped public attention. In the 1910s an attempt had been made by the Southern Pacific Railroad to market the forest as a tourist destination complete with a tourist camp. The railroad's booklets called it the "California Grove," a name that did not take.[15] That effort soon faded away, and by the mid-1920s, when Willard Van Name took up the question of its future, the grove had again faded into obscurity.

An irascible zoologist associated with the American Museum of Natural History in New York City, Van Name had been a critic of the political compromise struck in 1926 to enlarge Sequoia National Park. In a pamphlet published in 1927 he argued that the Park Service and the Forest Service had conspired when they established their new boundaries for park and forest so as to facilitate the logging of Redwood Mountain.[16] Nothing immediate came of Van Name's broadside, but over the next decade a general consensus grew that the grove did deserve protection. Critically, key property owners within the grove supported this direction, no doubt believing that in such poor economic times they were more likely to profit from selling to the government than from logging.[17] In anticipation of federal protection of the grove, the National Park Service made major purchases in the Redwood Mountain Grove in 1938, acquiring 3,720 acres from multiple owners at a total cost of $655,750.[18] Obviously, by paying an average of over $175 per acre, the government was offering a substantial premium over what it had expended to purchase the partially cutover lands in Converse Basin and east of Hume Lake.

Fortunately, over the winter of 1939/40, a Kings Canyon National Park bill was making its way through Congress. The legislation, written by Congressman Bud Gearhart of Fresno, included not only the creation of a large national park in the High Sierra but also authorized the president, by proclamation, to add up to 10,000 acres in the Redwood Canyon area to the new park.[19] In a statement before the House Committee on the Public Lands in March 1939, Interior Secretary Harold Ickes summarized the argument for the addition:

> This grove is generally conceded to be the finest remaining in private ownership and subject to exploitation. It is threated with destruction, not because the owners want to cut it, but because they cannot help themselves. They must sell the property or lose it, and it is hoped that they can sell it to the government. The bill provides that this tract may be added to the park by proclamation of the President.[20]

Gearhart's bill was approved by the full House of Representatives in July and then sent to the Senate. There it was carried over into the new year, where it passed on February 19 and was signed into law by President Franklin Roosevelt on March 4, 1940. Included in the final bill was a section that abolished General Grant National Park and transferred its lands to the new Kings Canyon National Park. Three months later, on June 21, Roosevelt issued a proclamation adding Redwood Canyon to the new park as well. Yet another large tract of giant sequoia forest had been returned to public ownership.

And that was not all that was going on in the late 1930s. To the south of Sequoia National Park, in the Tule River watershed, the fate of still another large private tract of Big Tree forest became an issue, and with some of the same individuals involved. Thomas Hume, in the years when he was organizing the Hume-Bennett Lumber Company and facilitating the transfer of operations of that endeavor from Converse Basin to Hume Lake, also invested in giant sequoia lands in the Tule River country. In 1905, the same year the Hume-Bennett Company took over Converse Basin, Hume purchased about 2,500 acres of giant sequoia land in the Mountain Home Grove, the grandest giant sequoia area in the Tule River watershed. During the next fifteen years, before the collapse of his Hume Lake operation, Hume continued his Tule River land purchases. Eventually he controlled more than 4,500 acres in the area, title to which Hume assigned to the Michigan Trust Company, apparently another Hume family operation.[21]

Hume's Mountain Home tract, like his forestlands around Hume Lake, all fell within the boundaries of the Sequoia National Forest, and the same negotiations that led to the sale of the northern lands to the Forest Service also discussed Hume's Mountain Home holdings. Even as the two parties worked out an agreement for the transfer of the mostly cutover northern lands, however, they could not agree on a price for the largely untouched Tule River acreage. During the 1920s, George Hume (Thomas Hume's son) initially allowed no harvesting of timber products to occur at Mountain Home, and although, beginning in 1929, contractors were allowed to cut up already-fallen sequoia logs for fence posts, the prohibition on cutting standing trees continued. After the collapse of purchase talks with the Forest Service, however, this policy changed. Beginning in 1941, a contract mill was allowed to establish itself at Mountain Home, with the apparent goal of threatening the trees as a means of leading to government purchase. For the first time in more than a generation, the sound of giant sequoias crashing to the earth again resonated through the Mountain Home Grove.[22]

At this point, Fresno attorney Arthur H. Drew stepped into the picture. A long-time friend of both George Hume as well as Hume's Fresno attorney Strother Walton, Drew had undoubtedly followed Hume's efforts to sell all his giant sequoia holdings to the Forest Service. Now, with that option apparently an unlikely future for the Mountain Home Grove, Drew decided to mount an effort to bring Hume's Tule River Big Trees into ownership by the State of California. Utilizing his status as an officer in the Fresno Parlor of the Native Sons of the Golden West, Drew began a campaign to appropriate state funds to purchase the grove.

With the support of state senators from both Fresno and Tulare Counties, a bill was introduced in 1943. It failed in its first attempt, but when it was reintroduced in the next session of the state legislature, it did better. Both houses passed the bill in June 1945 and it went to the governor for signature. Despite this legislative success, the State Board of Forestry remained unenthused about the potential purchase and recommended that Governor Earl Warren veto the bill. The board believed that other state priorities should take precedence. Instead of following the board's recommendation, however, Governor Warren sent State Forester DeWitt Nelson down to Tulare County to look at the grove and bring him a recommendation. When Nelson arrived a few days later, he found logging operations vigorously continuing and Big Trees falling to dynamite. Concluding that the purposefully destructive logging must be stopped, Nelson decided to ignore the concerns of his own staff. He went back to Sacramento and advised Warren to sign the bill, which he did on July 19. Two days later, the Michigan Trust Company gave a thirty-day termination notice to the contract lumbermen active in the grove and presented the state with a ninety-day purchase option. The state paid $550,000 for the 4,560 acres.[23]

Bowing to the wishes of their constituents, the local men who drafted the Mountain Home legislation created not a state park like Calaveras Big Trees but rather a state forest. A key difference between the two was that, unlike a state park, the Mountain Home Demonstration State Forest would continue to pay local property taxes.[24] Another difference was that the state would continue to harvest trees at Mountain Home, but under very different rules than those set by the Michigan Trust Company. During 1946, the state's first year on the ground at Mountain Home, basic directions fell into place, and by July, much had been worked out, at least on paper. A "Work Plan and Program" for the demonstration forest would pursue a handful of basic goals. Both mature and young giant sequoias would be protected, but already downed Big Trees could be utilized. At the same time, the forest's other timber resources would be subject to harvest, but only on a selective basis that would emphasize reducing fire risk and removing the oldest trees. Facilities would also be constructed to support public camping and enjoyment of the area. The first timber sale was for 797 cords of already felled giant sequoia wood.[25]

For the next few years, the State of California continued to keep close watch on the fate of the state's remaining privately held groves of giant sequoias. Several factors encouraged this attention, but at the top of the list was a continuing political campaign to acquire the South Calaveras Grove. In the years immediately following the completion of the Second World War, the southern of the two Calaveras Big Tree groves remained in the hands of the Pickering Lumber Company, which was doing its best to meet the pent-up postwar demand for lumber. By 1949, Pickering had extended its logging railroad system into the immediate vicinity of the grove, and it would, in fact, have gone through the Big Tree area had it not been pressured to avoid the sequoias by the State Park Commission.[26]

So began a political confrontation that would run for nearly a decade. State acquisition of the South Grove at Calaveras proved difficult for several reasons. The low prices for timberlands that had typified the 1930s were now gone. The postwar housing boom had all the nation's timber producers working at full output, and Pickering had no reason to sell cheap. Knowing that its lands in and around the South Grove contained dense stands of marketable pine and fir, the company held out for top dollar. Making the purchase even more difficult was state law that required that lands purchased for state parks be bought under a matching-funds formula that required substantial outside contributions. Such money, proponents of enlargement soon realized, was not easy to raise in light of Pickering's multimillion-dollar asking price.[27]

There was also the nagging question of whether, even if the funds could be obtained, the South Grove at Calaveras would be the best Big Tree purchase. In 1951, attempting to gain perspective on this question, the California State Senate instructed that the state's department of natural resources study all the privately held giant sequoia lands in the Sierra

Nevada and make recommendations as to which ones, if any, might deserve to be preserved.[28] Adding urgency to this request was the fact that giant sequoia logging was once again on the upswing, particularly in the Tule River country of Tulare County. The owners of the Dillonwood Grove, a large sequoia forest located on the North Fork of the Tule River just outside Sequoia National Park, had resumed logging there in 1948. Many large trees were felled during the next several seasons, and a photograph of this messy logging was published in the October 1951 issue of *National Geographic*.[29] The timing was no accident; the magazine was doing everything it could to generate public support for the acquisition of the South Grove at Calaveras.[30] In his article "Saving the Earth's Oldest Living Things," author Andrew Brown recalled the sequoia's many superlative qualities and recommitted the National Geographic Society to continued advocacy for their protection.[31]

The state report, which came out in late 1952, provided the first full overview of the status of the Big Trees since the US Department of Agriculture had drafted its report on the same subject for Congress fifty-two years earlier.[32] Both, coincidentally, had been prompted by proposals to acquire Big Tree lands at Calaveras. In its summary, the state provided a clear overview of who owned the Sierra's Big Trees and the threats faced by the individual groves. As a result of much tedious research, the report could now document that the Big Trees of the Sierra, always growing with other species present, could be found on about 35,600 acres. Of this land, private parties owned about 12 percent; the remainder was under some sort of public ownership, mostly national park (41 percent) or national forest (38 percent). The State of California controlled another 7 percent at Calaveras Big Trees State Park and Mountain Home Demonstration State Forest. The report also documented that about 23,500 of the 35,600 acres of giant sequoia land in the Sierra Nevada remained in a "virgin" (unlogged) state. Only 8 percent of this uncut land was in private hands, and except for the South Grove at Calaveras, all the remaining private holdings of this sort were in Tulare County.[33] Using the same data, the document estimated that "early logging" between 1870 and 1910 had consumed about 21 percent of the original forest.[34] A large percentage of this cutting, of course, had occurred in Converse Basin and the Hume-Bennett lands east and north of Hume Lake—land that had now come back into public ownership as a part of the Sequoia National Forest.

In its recommendations, the 1952 report identified three groves with exceptional Big Tree stands that possessed a high likelihood of being logged: the South Grove at Calaveras, and the Black Mountain and Freeman Creek Groves east of Porterville in the Sequoia National Forest. Each, the report's authors believed, would be a strong addition to the state park system. The report did not recommend acquiring Dillonwood.[35]

With the report in hand, the effort to acquire the South Grove accelerated. The federal government stepped in to help and transferred to the state a road corridor that would connect the two Calaveras groves. A decision that this donation could be counted as a part of the matching-funds balance needed to consummate the purchase strengthened the state's position. Quiet politicking by Newton Drury, director of state parks, and Horace Albright, retired director of the National Park Service, brought in additional funds in the form of a $1 million commitment from John D. Rockefeller, Jr. Sensing the inevitable, the Pickering Lumber Company negotiated aggressively to protect its financial interests. Finally, after a long series of offers and counteroffers, a compromise was reached in April 1954. Pickering would receive $2.8 million for 2,155 acres in and around the South Grove. It would also have the right to selectively log some of these lands prior to turning them over to the state.[36] The famous sequoias, of course, were off-limits. The South Grove at Calaveras finally belonged to the people of California.

All this activity reflected a key point: To a growing majority of citizens, cutting giant sequoias simply felt wrong. These ancient and compelling forest titans deserved better. Out of this flowed a continuing determination to, grove by grove, recapture the best surviving stands and groves that had been sold in the nineteenth century and return them to public ownership. Eventually, not only the National Park Service but also the US Forest Service and the State of California were all affected. Even in the timber boom that followed the Second World War, public support for protecting the Big Trees endured.

A legal opinion in 1951 demonstrated just how strongly Californians had come to feel about their Big Trees. In 1937 the state legislature had designated the "California redwood" as the official tree of the Golden State. Inevitably, this vague designation led to arguments as to whether this only applied to *Sequoia sempervirens* (the redwoods growing in northern California along the coast) or also to what was then still known as *Sequoia gigantea* — the Big Trees. Stepping in to resolve this issue, Attorney General Edmund G. "Pat" Brown, Sr., issued an opinion in 1951 that the two species shared the title. The Big Trees were now not only too special to cut, they were also an official symbol of the State of California.[37]

CHAPTER FIFTEEN
Kindled Light

I've been strolling for the past hour along one of my favorite Big Tree trails. My route has brought me down into the creek-side portion of the Redwood Mountain Grove. It's easy enough to see why Franklin Roosevelt used the authority Congress gave him in 1940 to add these lands to Kings Canyon National Park. From the time I left my car at the trailhead, I have been walking through a seemingly endless giant sequoia forest. This is, after all, the largest of all the groves. But that's not why I have come today.

After several miles I arrive at a junction. Taking the right-hand fork, I begin climbing steeply. I leave behind the dense stand of Big Trees that prospers along Redwood Creek and ascend through a drier and thinner forest. Sequoias persist here too, but not with the same density found in the moist canyon bottom. Sweating now, I eventually approach my destination. I've been climbing through brushy stands of ceanothus and manzanita, shrubs that seldom rise much taller than my head, but suddenly the vegetation changes. I enter an extensive thicket of sapling trees, a conifer jungle so thick I can barely see a dozen feet into it.

I hardly need to look at the blue-green foliage to know that I have pushed into a dense stand of young giant sequoias. Narrow, spire-topped trees rise all around me, competing with each other for the sunlight they need to grow. It's not hard to separate the winners from the losers in this game of life and death. Dead, sun-starved tree skeletons surround me. Above them tower more successful trees, many of them now thirty or more feet tall.

Big sequoias usually grow with pines and firs in mixed-species stands, but this young forest allows for none of that. Around me I see nothing but sapling sequoias—thousands of them. I know the story of this place. In 1977 national park fire managers ignited a fire here—a "prescribed fire," to use the government's term. On this steep slope the resulting blaze burned intensely, and by the time it had run its course the only trees left alive here were a handful of adult sequoias, trees tall enough to rise above the fire that consumed their smaller and shorter neighbors. A wet winter with heavy snows followed, and when spring arrived, this mountainside of ashes came to life. Countless giant sequoias sprouted—hundreds per square yard in some places. Each emerged from a seed no larger than a flake of rolled oats—a seed that had rained down from the surviving large sequoias in the aftermath of the previous year's fire.

Rangers lighting forest fires to stimulate sequoia reproduction! In 1977 the idea still seemed radical. It had come into focus only in the previous decade, and even after the science confirmed the methodology, many thought the concept crazy. But around me, as I stand on the slopes of Redwood Mountain, the effectiveness of this strategy is hard to argue. How can one ignore such obvious success?

As noted earlier, Native Americans often made extensive use of fire in the forests of California, including in the giant sequoia groves of the Sierra Nevada, but this proactive attitude largely disappeared in the face of Euro-American attitudes about fire. From Muir on, most students of the Big Trees saw fire as the species' most destructive enemy. Muir said as much in his 1894 book, *The Mountains of California*, when he called fire "the great destroyer of Sequoia." The naturalist had passed his formative years in moist environments where fire did not play a significant role, and he accepted the popular equation that trees bearing scars had suffered significant damage. Since all large Big Trees displayed such scars, and badly burned dead snags could be found in many groves, the threat of fire was demonstrably real. As he wandered through the groves, however, Muir paid sufficient attention to notice that there was more to the story than simple damage. He noted that fires furnished "bare virgin ground, one of the conditions essential for [the trees'] growth from seed." Muir had stumbled onto something important here, but he underestimated the significance of what he had realized when he remarked, "Fresh ground is, however, provided in sufficient quantities for the constant renewal of the forest without fire, viz., by the fall of old trees."[1] In this, it would turn out, Muir was simply wrong.

Over time, others discerned some of the same significant points. In 1908, Forest Service dendrologist George B. Sudworth, in his monumental *Forest Trees of the Pacific Slope*, captured several of the Big Trees' key attributes:

> General absence of reproduction in all but openings in forest and in open ground adjacent to seed trees shows clearly that light is the most important factor in early life.[2]

And later on the same page:

> Reproduction generally best on burned areas, where fire has cleared off litter, and exposed mineral soil, or even after light ground fire has left a layer of ashes or charcoal.[3]

Sudworth, who would cap his career with a term as the chief dendrologist for the US Forest Service, was an astute observer of trees, and he was on the threshold of understanding the Big Trees in a fundamentally new way, but he never quite got there. Neither did most of his peers. What happened instead was that, with the Forest Service strongly in the lead, both land managers and scientists bought into the intellectual paradigm that wildland fires of all sorts and under all conditions were negative events that needed to be suppressed. The possibility that fire might actually benefit some species was rejected out of hand.

The "demon fire" perspectives arose out of forestry practices in humid northern Europe, where Germans had worked out the basics of modern forest management in their heavily utilized woods. Gifford Pinchot, the first certified forester to practice in the United States, had been trained in this perspective, and when he assumed control of the nation's national forests in 1905, he began the systematic application of this idea. Strengthening its acceptance was an idea held strongly by the Progressives, a political movement of the time that believed strongly in the efficient utilization of natural resources.

To the Progressives, natural resources were intended for human use, and the primary objective was their intelligent management. As early as 1899, six years before he became the first chief of the US Forest Service, Pinchot wrote a two-volume work he titled *A Primer of Forestry*. The section that addressed fire fell within the chapter "Enemies of the Forest," and the opening sentence regarding fire clearly summed up Pinchot's position: "Of all the foes which attack the woodlands of North America no other is so terrible as fire."[4] The pages that followed provided both examples of historic blazes that did great damage as well as techniques for controlling fire. By 1907, when Pinchot issued his first manual for field personnel in the newly organized US Forest Service, the assumption that all fires must be suppressed was deeply embedded throughout the slim volume. He contrasted Forest Service management with "the old condition on the public domain, where fires were incessant and enormously destructive."[5]

Pinchot left the Forest Service in 1910, fired by President Taft for arguing with the secretary of the interior over federal land management policies, but he had trained the men who succeeded him, and they ensured that his attitudes prevailed. Then, only a few months after Pinchot's dismissal, huge fires swept over the northern Rocky Mountains. Three million acres burned, at least eighty-seven persons died, and smoke drifted all the way to the Atlantic seaboard. Out of this searing experience came a Forest Service even more committed to the total suppression of fire.[6] For decades, the possibility that wildland fires might sometimes have beneficial effects simply would not be seriously discussed.

None of the giant sequoia literature that came out in the years after 1910 challenged this position. What *was* safe to write about was how resilient the Big Trees were when challenged by fire. In his chapter in the 1921 *Handbook of Yosemite National Park*, Jepson wrote about their thick bark and lack of flammable resin.[7] Three years later, in *The Giant Sequoia*, Rodney Sydes Ellsworth limited his remarks about fire and the Big Trees to comments that the tree's ability to heal itself after being scarred was a monument to its "vitality."[8] Fry and White were equally cautious in *Big Trees*. In a single sentence offered without elaboration, they noted, "Although forest fires are the most deadly enemy of the grown tree, they play an important role in promoting the reproduction of new trees."[9] Beyond that, they, like nearly everyone else, emphasized the ability of the trees to resist fire. Discussing a Big Tree as if it were a person, Fry and White wrote:

> He lives through fire after fire. The debris-littered forest may flame around him for hours or for days; scores or hundreds of fires may flicker and roar through the centuries over his thick asbestos bark, but almost in vain; nine-tenths or more of that bark may finally be charred away; two hundred feet of his vitals may be eaten out so that he stands a mere shell—a chimney or window tree—and yet he lives on![10]

Fire remained, in book after book, the "enemy" of the giant sequoia. Behind the scenes, however, national park managers were beginning to ask questions. They were not initially focused on the role of fire, but eventually that is where their inquiries would lead. The immediate problem, as national park managers at Yosemite saw it, was that monarch sequoias continued to fall down. In the Mariposa Grove, the Massachusetts Tree had crashed to the ground in 1927; the Utah Tree and the Stable Tree followed suit in the winter of 1933/34. The Mark Twain Tree collapsed in 1943, to be followed a year later by the San Francisco Tree. An unnamed giant fell during the winter of 1949, and the Iowa Tree, a well-known specimen within sight of the Big Trees Lodge, came down in February 1954.[11]

With Meinecke's recommendations still on their minds, park staff also noticed other problems, such as soil erosion in the immediate vicinity of the Wawona Tunnel Tree. Park staff even found themselves pondering whether the sequoias were becoming a safety hazard to visitors. In the summer of 1954, the park responded with several small initiatives. To stabilize the Tunnel Tree, park crews constructed a rock retaining wall to limit erosion that was threatening the tree. The park also organized an internal committee to study "sequoia problems." The committee included both the park forester and the park's chief naturalist, the closest position Yosemite then had to a scientist.[12] Eventually, the committee worked up a list of recommendations about the management of the Mariposa Grove that were intended to mitigate the visitor impacts that the committee found so obvious. Item number one on the list was to employ the services of "trained forest ecologists" to survey the negative effects of heavy visitation.[13]

The need to find a way to carry out visitor-impact research within the Mariposa Grove fell to chief park naturalist Douglass Hubbard. Working closely with superintendent John Preston and regional office staff including biologist Lowell Sumner, Hubbard sought funds and organized a study. He also found someone to carry it out: a young park naturalist named Richard J. Hartesveldt. Born in 1921, Hartesveldt found himself attracted to national park work at an early age. By 1952 he held a position as a seasonal ranger-naturalist at Devils Postpile National Monument, and by 1955 he had advanced to the position of Senior Ranger Naturalist in Yosemite. Seeking to further advance his career in an agency that still saw ranger-naturalists as scientists, Hartesveldt began to consider the pursuit of a doctoral degree. Exploring this idea with Hubbard, the two men worked out a plan that involved Hartesveldt pursuing his degree at the University of Michigan with a dissertation focused on human impacts in the Mariposa Grove.

The decision to select the graduate program at Michigan was no accident. There, Hartesveldt became a student of Stanley Cain. Born in Indiana in 1902, Cain had begun teaching botany in the 1920s and had gone on to earn both MA and PhD degrees in that subject from the University of Chicago. Taking a position at the University of Tennessee, Cain's interests continued to broaden, and he wrote about plant geography and began to explore questions in the new field of ecology. He also began to think hard about questions related to the conservation of natural resources. In 1950, Cain was recruited by the University of Michigan, where he became the founding chairman of the nation's first university-level department of conservation. This was the program Richard Hartesveldt joined as a graduate student.

Assisted now by his academic advisors, Hartesveldt worked out the basic questions that needed to be answered. Central to his work would be "the simple objective of determining man's influence on *Sequoia gigantea* and to discover any modifications of the normal life processes of these trees which, because of the influence of man, might prove to be detrimental to their health and vigor."[14] From the beginning, this meant that Hartesveldt had to devise a means for quantifying impact. As the study progressed, he also defined two additional goals, which were to examine past management practices and their results and to propose new management policies and procedures to protect the trees.

Field research began in the summer of 1957 on a part-time basis while Hartesveldt continued to carry out his ranger-naturalist duties. The work continued in this manner through the summer of 1958 but then was accelerated in 1959 when Superintendent Preston came up with funding for a summer research position that allowed Hartesveldt to work full-time on his field studies. Fieldwork ended in September 1959, and by November 1961 Hartesveldt had completed his analyses and drafted his recommendations. Helping support all of this were grants from the National Science Foundation and the National Wildlife Federation.[15]

Hartesveldt's dissertation can be seen as a key turning point in thought about not only the management of giant sequoias but also about national park policy informing the management of all natural resources. After laying out the human history of the Mariposa Grove and the many things that had been done to it, Hartesveldt moved into the heart of his study with a description of several types of human effects. These included direct mechanical attacks on individual trees (issues varying from vandalism to road construction) as well as impacts from the "multitudinous repetition" of smaller injuries, by which he meant primarily soil compaction and erosion. To study these impacts, Hartesveldt focused both on the health of the trees' root systems and possible resulting changes in tree vigor (i.e., growth rates).[16] So far, Hartesveldt was following closely in the footsteps first laid down nearly thirty years earlier by Emilio Meinecke. But there was more.

In seeking to document the status of a certain plant species growing in the meadow adjoining the grove museum (the two-room log cabin built in 1930 on the site of Clark's original "hospice"), Hartesveldt sought out a set of detailed photos that had been taken of the grove in 1935. What he discovered surprised the naturalist. In the quarter century that had passed since the photos were made, the visible appearance of the grove had changed significantly. A meadow had popped up in a previously dry locale, and thickets of young fir trees had filled in previously open areas within the Big Trees stands. These insights sent Hartesveldt into the park's historical photos files in search of other images that might shed light on changes within the grove. Soon he had amassed a collection of two hundred such images. In the summers of 1958 and 1959, he began systematically rephotographing these scenes with the goal of documenting changes. He also resurveyed a portion of the grove that was the subject of a very detailed plant map that had been made in 1934.[17]

Working closely with his academic advisors from the University of Michigan, Hartesveldt now realized that he had discovered an entirely new and previously unstudied class of human impacts upon the forest. These impacts, he realized, were not mechanical but rather ecological, in that they related to how the grove's myriad life-forms interacted with each other and to natural processes like moisture flow and fire. Soon, Hartesveldt was able to identify several broad patterns of ecological change within the grove. Some, he discerned, were wildlife related—issues like intense browsing by large deer populations. Another major change was the accidental creation of a wet meadow near the museum by overflow spillage from the water system installed in 1931. Among other things, this addition of what Hartesveldt estimated to be over seven million gallons of water each year to the grove might have led to the collapse of several sequoias.[18] But by far the most significant ecological impact Hartesveldt identified in the grove related to changes in plant succession.

Like every other forest in the Sierra Nevada, the Mariposa Grove consisted of a mix of plants with varying needs. Some grew well in dense shade while others required bright sunshine. In general, Sierra forests moved back and forth between these conditions as a result of the periodic intrusion of fire. It was this sporadic disturbance that allowed sun-loving plants, including sequoias, to germinate and grow in forests that tended otherwise to move over time toward dominance by shade-tolerant species. All this had been known for some time, at least by foresters and silviculturalists, but in the Sierra few had given much thought to what might happen when forests were subjected to prolonged fire suppression. Now, by studying his comparative photographs, Hartesveldt could see what was happening.

In what would turn out to be by far the most important of his findings, Hartesveldt proposed that fire suppression had allowed the Mariposa Grove, previously a relatively open forest kept that way by fires, to fill in with thickets of shade-tolerant white fir trees. Looking back through the many words written about the grove since it was first set aside in 1864, he could find little significant appreciation of what fire suppression was doing to change the grove and its forest. In the absence of fire, which had apparently occurred repeatedly in the grove in times past, as evidenced by numerous fire scars on the Big Trees, the forest was moving through a successional process that changed its nature profoundly. The resulting "new" forest provided only minimal opportunities for sequoia reproduction, threatened established sequoias by increasing fire risk (through fuel accumulation), and even interfered with human enjoyment of the grove by obscuring views of the very trees visitors came to see.[19] Here was ecological change on a previously unimagined scale.

As expected, Hartesveldt's doctoral dissertation also provided the managers of Yosemite National Park with a substantial list of recommendations for the future management of the Mariposa Grove. Unsurprisingly, many of these grew directly from the various forms of direct mechanical damage that Hartesveldt had identified as occurring within the grove. For starters, the Park Service needed to pay much more attention to protecting giant sequoia roots, an effort that would affect everything from underground utility lines to roads and parking lots. Also, the accidental wet meadow needed to be phased out. On a broader scale, the Park Service would have to shift its attention from managing to protect individual Big Trees to a wider ecological approach that placed primary value on ensuring that the Mariposa Grove was a healthy and sustainable *forest community*.[20] Such an approach, of course, would require a thorough ecological understanding of the giant sequoias, and after three summers of research in the Mariposa Grove, Hartesveldt now knew how little he, or anyone else, knew about how giant sequoia forests actually worked. His closing recommendation was that the National Park Service undertake further research. High on his list of things to be studied was the biggest issue he had uncovered: what to do about the fires. The Park Service needed to "study plant successional trends in the absence of wildfire and human influences and evaluate the effects upon the health and growth rate of the sequoia trees."[21] He may not have yet have fully understood it, but this challenge would become the primary focus of the remainder of Hartesveldt's professional life.

Shortly after earning his doctorate, Richard Hartesveldt resigned from the National Park Service and accepted a position as an associate professor of conservation at San Jose State College in California.[22] His links to the Park Service remained strong, however, and by the summer of 1964 he was hard at work pursuing the very research he had proposed in his dissertation. Hartesveldt was back among the Big Trees, but this new project was no longer taking place within the Mariposa Grove of Yosemite National Park. Instead, the Park Service had decided to pursue its giant sequoia ecological research in a quieter and less visible location: the Redwood Mountain Grove of Kings Canyon National Park.[23] Funded now by the National Park Service, Hartesveldt assembled a research team composed of fellow San Jose State colleagues. H. Thomas Harvey joined Hartesveldt in his tree ecology studies, and Howard Shellhammer came on board to study the relationship between vertebrate species and the sequoias. In 1966, Ronald E. Stecker also joined the team to study insect issues. Within the Redwood Canyon section of the Redwood Mountain Grove, the San Jose State team established four research plots and began experiments at each. In 1965 and 1966 they very cautiously burned small portions of the plots; it was the first experimental burning for ecological purposes to take place within a national park in the western United States.[24]

Although the San Jose team was breaking new ground, they were working within a broader context of change that did much to allow them to pursue their studies. At the University of California, Berkeley, Harold Biswell was carrying out continuing research about the natural role of fire in California's forests. Biswell had joined the UC staff in 1947 and soon was doing purposeful burning in yellow pine forests in the central Sierra. This work eventually attracted more than just scientific attention. The 1959 annual issue of the *Sierra Club Bulletin* contained his essay "Man and Fire in Ponderosa Pine in the Sierra Nevada of California," in which he clearly laid out a case for the beneficial effects of fire.[25] Two years later, extending his attention to the sequoias, Biswell wrote an article for the magazine *National Parks* titled "Big Trees and Fire," and here he made a case for the importance of fire in preserving healthy giant sequoia forest communities.²⁶

Hartesveldt, who was just finishing up his dissertation when this came out, cited Biswell in his bibliography. In 1964, Biswell initiated experimental prescribed burning among the Big Trees at Whitaker's Forest, a 320-acre research tract owned by the University of California.²⁷ Whitaker's adjoined Kings Canyon National Park, and Biswell's burn sites were only a mile or two from those where Hartesveldt and his team began work the following year. Biswell continued fire research at Whitaker's for the next decade and attracted many eminent researchers to the site for workshops.²⁸

At the same time, the Park Service was wrestling on the national level with the challenge of integrating ecological principles into park management policies, and the agency continued to be criticized as being focused too heavily on tourism and visitor facilities. Out of this situation arose what came to be known as the Leopold Report, more formally "Wildlife Management in the National Parks." Professor A. Starker Leopold of the University of California, Berkeley, chaired the committee that wrote the report and served as its primary author.²⁹ Despite its title, the document actually addressed ecological management in national parks as a whole, and it outlined what amounted to a new goal for the agency. "Unimpaired" national parks, it said, should henceforth be defined in terms of how closely they matched the condition of landscapes *and ecosystems* before they were modified by the actions of Euro-Americans.³⁰ Ecological thought dominated this new approach, and the role of fire was not ignored. The report advocated the controlled use of fire to achieve ecological goals, and it specifically mentioned that the giant sequoia trees of Yosemite, Sequoia, and Kings Canyon were at risk from the accumulation of excess levels of forest fuels. The situation required "immense concern" on the part of park managers.³¹

Biswell's continuing work on fire at Whitaker's Forest combined with the Leopold Report to strengthen the importance of the research being done by the San Jose State team in Kings Canyon National Park. Despite the presence of many traditionalists within the Park Service—men who had grown up with and believed in total fire suppression—the tide began to turn. In 1968, the Park Service hired Bruce Kilgore and assigned him to

Sequoia and Kings Canyon. Kilgore had grown up camping in the Sierra with his family and earned a bachelor's degree in conservation at the University of California in 1952. After that, he went on to assignments as the editor of *National Parks*, and then, by 1960, the *Sierra Club Bulletin*, where he worked closely with David Brower, the club's charismatic executive director. In 1963, under Kilgore's editorial direction, the *Bulletin* published the Leopold Report in its entirety, and excited by the importance of these changes, Kilgore left the Sierra Club and began work on a doctoral degree in zoology under A. Starker Leopold. In 1967, as Kilgore completed his doctoral research, Leopold lobbied the Park Service to assign him to Sequoia and Kings Canyon as the twin parks' first research scientist. There, Kilgore quickly established himself as a leading advocate for prescribed fire.[32] In 1968, the same year that Kilgore came aboard at Sequoia and Kings Canyon, the parks committed to a long-term program of fire research and prescribed burning.[33] Yosemite followed suit two years later. Implementation would take time, but fire had returned to resume its natural role in the sequoia groves of the three Sierra Nevada national parks.

The San Jose State team went on to produce three books about their work. The first, *The Giant Sequoia of the Sierra Nevada*, was completed in 1971, although it was not published by the government until late 1975, shortly after Hartesveldt's untimely death from a heart attack.[34] The three remaining members of the team, now led by Tom Harvey, continued their giant sequoia work, and out of their efforts came a second, more technical book in 1980.[35] The following year, the Sequoia Natural History Association brought out a popular book by the team that condensed the best of the other two works and has remained in print for more than thirty years as the standard popular reference about the Big Trees.[36]

Half a century later, we can see that the giant sequoias played a starring role in the redefining of America's relationship with wildland fire. For decades, national consensus had defended a policy of complete fire suppression, and all fire was perceived as an enemy of trees, including sequoias. Earlier attempts had been made to crack open this prevailing attitude, but it took the imagination of men such as Biswell and Hartesveldt, as well as the majesty of the Big Trees of the Mariposa Grove, to bring real change.

Hartesveldt's doctoral studies at Yosemite and his follow-up work with the San Jose State team at Kings Canyon produced so compelling a story that the beneficial effects of fire under certain conditions could no longer be ignored. After 1970, at least in the three Sierra Nevada national parks, fire became the preferred tool for maintaining natural conditions in sequoia groves, and in response, a new generation of sequoia seedlings sprouted. To the descendants of the native peoples who had lived in the groves for so long, the change confirmed a return to long-overdue common sense.

That studies focusing only on giant sequoias were capable of changing public attitudes about much larger issues demonstrated how far the sequoias had come. Now, it seemed, they were not just worthy in their own right but also carried the distinction of being symbols for the entire natural world. If fire was a necessary component of sequoia forests, then might it play beneficial roles in other ecosystems as well? With the Big Trees as a driving force, the long reign of total fire suppression in the United States' protected natural areas began to crumble, and a new set of more ecologically based attitudes arose to replace the doctrines that had dominated forest management for so long.

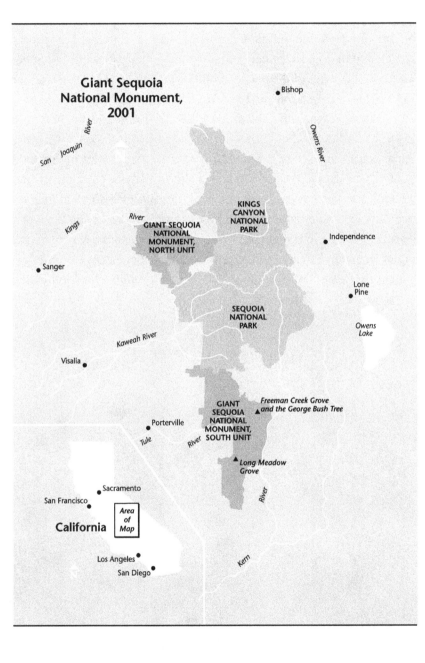

CHAPTER SIXTEEN
Worth the Fight

At first glance this forest looks to be anything but controversial. Visitors stroll through the sequoias on a paved path, taking photos and marveling at the towering giants. I hear English spoken with a bit of a Southern drawl—perhaps a family from nearby Bakersfield with its strong Dust Bowl roots. Another group speaks German; yet another chatters about the scene in Spanish in an accent that suggests Mexican roots. I see people everywhere. This forest trail is popular by any measure.

A family pushing a baby stroller passes me. A few dozen yards ahead I see them pause and pick up the child. Mother and daughter pose for a photo in front of a huge red tree, creating a compelling image that contrasts youth and immense age, fragility and enormous resilience. I pause too, stopping to peruse the displays that line the trail. A quick reading confirms that I am in a national forest, not a national park; the message here is not one of preservation for preservation's sake. One panel mentions that a ponderosa pine along the trail is "highly desirable for a number of wood uses including paneling, siding, studs and decking." Another panel, standing close to a fallen sequoia, asks, "Should these fallen giants be harvested and used to meet man's needs, or preserved (as this is) for future generations to enjoy?" Judging from the condition of the exhibits, visitors have been inspired to answer the question right here. The display has been vandalized, and the line asking about cutting up fallen trees has been almost obliterated. Even an exhibit that offers nothing more controversial than the observation that "in their stately grandeur [the trees] carry the wisdom of time" had been attacked; someone using a sharp object has changed the title of the panel from "Reflection Pool" to "Erosion Pool," apparently noting that whatever pool once stood here has now filled in with silt.

The "Trail of a Hundred Giants" occupies a place of honor in the Long Meadow Grove of Big Trees. I've come, at last, to very near the southern end of the giant sequoia belt. Here, on the Sequoia National Forest of Tulare County, the trees extend beyond the Tule River country and grow in a handful of locations that drain eastward into the Kern River watershed. At Long Meadow, the Big Trees occupy an area of almost three hundred acres. No complete count has ever been made of sequoias here, but estimates run upward of two hundred large trees.[1] The grove, for those who explore it carefully, also possesses a dual identity that can only be described as schizophrenic. The portion of the grove that features the Trail of a Hundred Giants stands as the most publicized of all the sequoia groves in the Giant Sequoia National Monument, a reserve overlain onto the Sequoia National Forest in 2000 to protect old-growth sequoia forests. Yet, less than a mile away, the same grove contains acreage where the Forest Service oversaw intense logging as recently as thirty years ago.

It takes about half an hour and a good forest map to drive from the tourist side of the Long Meadow Grove to the portion managed not so long ago for timber production. A maze of mostly unmarked logging roads circles around to the western side of the grove. I pass a meadow full of cattle, continue another mile, and only then reenter the sequoias. Here, I find no parking areas, no paved paths, no interpretive signs. Dense, almost impenetrable thickets of brush and young conifer trees rise twenty or thirty feet into the air. Above them, standing like lonely skyscrapers, mature sequoias rise up two hundred feet into the cerulean sky. Here, on land that was first set aside in the 1890s to prevent its abuse, the Forest Service tried in the 1980s to bring together modern industrial forestry and giant sequoia trees. Within the Long Meadow Grove, about one-third of the Big Tree area was logged, including seventeen acres that were clear-cut of all non-sequoia trees. Other groves received similar treatment.² The result was a bitter multi-decade political battle over how the sequoia groves located on the Sequoia National Forest ought be managed. Along the way, a new national monument came into existence, and a generation of attorneys found work in a long series of lawsuits and appeals.

Before we can explore this prolonged battle, however, we must set the stage by telling another story—one about forest restoration in national parks.

In the aftermath of Richard Hartesveldt's work in the Mariposa Grove, the managers of Yosemite National Park wrestled with what they now knew to be undeniable human impacts within the grove. Hartesveldt's advice that the grove be managed to sustain the health of the entire forest rather than just the condition of individual trees led to a thinning and burning project beginning in 1969. But even as ecologically based forest management began in the grove, visitor-use problems continued. This forced park managers to contemplate more profound changes. As plans were being developed for the reintroduction of fire into the grove, park officials committed also to a program to remove private vehicles from the loop road that led past the Grizzly Giant and up to the famous Tunnel Tree. The new approach called for constructing a 130-space parking area among the Big Trees at the entrance to the grove and restricting use of the existing road system above that point to shuttle and tour buses. Construction was slated to begin in the spring of 1969.³

One of the goals of this change was to protect the fragile giants everyone wanted to see. High on the list, of course, was the Tunnel Tree, probably the best-known tree in the national park system. Unfortunately, the tree did not live to see the new regime designed to protect it. Weakened over the years by deep natural fire scars, by tunnel excavation in 1881, and by visitor-caused erosion, the 234-foot-high tree splintered apart and fell, probably during the series of intense snowstorms that swept the Sierra in February 1969.⁴ No one saw the tree come down, and its collapse went unnoticed until May, when a Park Service snowplow crew clearing the road found the famous giant on the ground.

Despite the loss of the Tunnel Tree, Yosemite persevered with its Mariposa Grove plans, and the 1970 season saw the grove's roads closed permanently to private automobiles. The shift to shuttle buses for grove access

The famous Wawona Tunnel Tree fractured and collapsed during a heavy snowstorm in March 1969. Putting tunnels through standing trees, it turned out, was not a good idea. COURTESY OF YOSEMITE NPS LIBRARY

placed the Big Trees Lodge in an awkward situation, however, and the Park Service closed the facility in 1972.[5] The concessioner, which might have been expected to object, apparently had found the twelve-room facility too small to be economical to operate and did not fight the closure. Together, the removal of automobiles and the closure of the Big Trees Lodge represented the end of the Mather/Albright era in the Mariposa Grove. It had lasted a full half-century.

In the Giant Forest of Sequoia National Park, the Park Service faced similar problems, but in several critical ways the challenge was far larger. Unlike the situation in the Mariposa Grove, Sequoia National Park's main highway passed through the center of the Giant Forest, and the lodges there offered not twelve rooms for rent but more than three hundred. Compounding the issue was the fact that the government had signed a twenty-year contract with the concessioner in 1952, which eliminated all hope of removing facilities from the grove during that period. Confirming this attitude was a decision made in 1950 to allow the felling of a leaning eighteen-foot-thick giant sequoia to protect rental cabins.[6]

The need for a new direction here, however, was even more compelling than it was in Yosemite. Park Service staff at Sequoia had not forgotten the 1927 Meinecke report, and they had followed with interest the ongoing discussions about the future of the Mariposa Grove. In 1960 the southern park commissioned a group of NPS officials to review long-term directions for the Giant Forest, and that group concluded that Superintendent White's plan to remove commercial development from the grove continued to make sense. Still, with a long-term concessions contract in place, the implied goal of removing lodging from the Giant Forest seemed unrealistic. What could not be ignored, however, was the traffic congestion problem that had grown to crisis proportions in the developed part of the grove. On many busy summer days, traffic simply came to a complete halt where the park highway passed through the Giant Forest Village area. All this came to a head in the late 1960s, when the Park Service began preparing a new master plan for Sequoia and Kings Canyon National Parks. The planning included congestion studies and explored several options for a bypass road that would take through traffic away from the heart of the giant sequoia area. None of this pleased the park concessioner, which earned the great majority of its income at the Giant Forest, and when the dust settled, the 1971 master plan chose not to threaten those facilities. Instead, the government removed most of its own services from the area, including three campgrounds, with a total of 152 popular campsites, and the Sequoia National Park post office. It also signed another twenty-year contract that left the concessioner's operations in place.

Recognizing, nevertheless, that something had to be done, the master plan did contain a recommendation that the Giant Forest area required at least a physical remodeling of its visitor infrastructure, and it was that effort that finally opened the door to change in the grove. The agency hired an outside planning firm to lead the effort, and it was this non-park team that pointed out the obvious: that the Giant Forest, the park's most visited feature, had not just traffic congestion problems but also visitor-experience and ecological problems. Big changes were in order. The outside design firm continued working on the Giant Forest problem, and out of these new studies came a draft plan in 1974 that called for the removal of all commercial facilities from the Big Tree area. The details took a while to work out, but when the final plan was approved in January 1980, it proposed a complete removal from the Big Tree area of all commercial facilities and services, and their subsequent relocation to nearby areas outside the grove.[7] It would take time to carry out the plan—over twenty years, in fact—but after 1980, the Park Service remained committed to reducing and eventually removing all commercial facilities from the Giant Forest. Not until late 1998 would the last of the lodgings close.

Eventually the Park Service would remove 282 buildings from the Giant Forest and over a million square feet of asphalt pavement. During these long years of incremental work, what continued to surprise park managers was the unflinching public support for the effort. Few visitors objected to the relocation of lodging and other services to less-sensitive sites within the park. In the mid-1990s, when the concessioner mounted a last-ditch effort to stave off final closure, the Park Service declared the company nonresponsive and terminated its preferential right of contract renewal. It had taken twenty years longer than at Yosemite, but the final closure of the Giant Forest Lodge marked the end of the Mather/Albright world in Sequoia National Park as well.[8] Henceforth, the prime visitor feature of Sequoia would be run as a day-use-only area managed specifically to protect the best of the Big Trees.

By the end of the 1990s, national park managers at both Yosemite and Sequoia had shifted their attitudes profoundly from those that the Park Service first applied when it assumed management of these areas in 1916. Gone was the expectation that individual specimen trees took precedence over all other natural concerns. Gone, too, was the perception that full visitor services must always be available within the groves. The agency had shifted to an ecological approach that put the health of the forest first, the fate of individual trees second, and the need for visitor services a distant third. By century's end, the Park Service was much more interested in perpetuating the role of fire in the groves than in ensuring the availability of postcards and ice cream bars. This huge shift in attitude, which the agency worked very hard to sell to the public through its interpretive programs and literature, resonated well with the great majority of park visitors. Public attitudes toward the sequoias had reached the point that the trees took precedence over nearly all other values. Earlier, this belief had fueled the continuing public acquisition of Big Tree lands, and now it had justified a fundamental reprioritization of National Park Service management goals. In the neighboring national forest, where roughly 40 percent of the sequoias were to be found, change took other directions—directions that would lead ultimately to prolonged political controversy.

In 1952, when the State of California prepared its report on the status of giant sequoias in the Sierra Nevada, that report interpreted Forest Service policy toward the Big Trees as ensuring that "the esthetic values are safeguarded." The report went on to state that it was "probable" that "on National Forest land, logging of all species may henceforth be excluded from the groves proper."[9] This attitude did not represent anything new on the part of the Forest Service, which up to this point had not allowed extensive timber harvest on the Sequoia National Forest, where the great majority of the agency's Big Trees grew. All this would change, however, over the next quarter century.

By the early 1970s, timber production had grown to become *the* dominant priority across the national forest system as a whole, and the Sequoia Forest, with its large expanses of coniferous woods, followed suit.[10] At the same time, however, local forest staff followed with interest the giant sequoia research taking place within the adjoining national park groves. The ecological insight that revealed that the Big Trees had for millennia relied upon natural fire to achieve successful reproduction was not a message that national forest managers received comfortably; since the time of Gifford Pinchot, the Forest Service had led the national effort to suppress all fire in wildland forests. Nevertheless, in 1975, in response to Park Service research, the Sequoia National Forest carried out a cautious experimental burn within the Bearskin Meadow Grove near Hume Lake. The fire's goals were to reduce flammable fuels and prepare seedbed areas for sequoia reproduction. In a post-fire analysis, however, the agency concluded that "neither objective was fully met."[11] The agency did not follow up with any additional fire experimentation because it had another idea, one more compatible with the saw-timber-production-program approach that the agency emphasized on other lands.

One lesson that the Forest Service had learned was that forest disturbance associated with intense logging could lead to vigorous giant sequoia reproduction. This phenomenon was particularly apparent in the cutover giant sequoia lands the agency had acquired in the 1930s. In places like

Converse Basin, which had seen the most extensive giant sequoia logging ever to occur, large stands of young giant sequoias now prospered. Could these young forests provide the agency with an ecological alternative to fire in its sequoia groves that might also facilitate timber harvest? The managers of the Sequoia National Forest thought so and promptly moved in that direction.

For five years, beginning in 1981, the Sequoia National Forest carried out timber sales in its Big Tree groves. During those years, thirteen sales were executed, covering about 1,000 acres. To quote the agency: "Approximately two-thirds of the cutting was designed to create conditions favorable for giant sequoia reproduction by clearing the forest floor and opening the canopy."[12] In practice, this meant felling and then either removing or burning all standing trees except specimen-sized giant sequoias. The resulting landscapes looked exactly like clear-cuts, which in essence they were, except for the presence of scattered, and often visibly lonely, monarch Big Trees. At the same time, the Forest Service also began an aggressive program of planting giant sequoias, both within the groves that were being logged and in other areas that had not naturally supported Big Tree growth. Some years, the agency planted as many as 40,000 seedling sequoias. By 1991, the agency had put 160,000 sequoia seedlings in the ground on about 1,400 acres. More than half of the planting took place outside historical grove boundaries.[13]

To Forest Service managers, this program of forest manipulation made clear and compelling sense. If giant sequoias were at risk from woody fuel accumulation and needed sunny, disturbed soils for successful germination, and if the nation placed a high premium on timber production, then why not pursue both goals simultaneously by harvesting "whitewoods" in sequoia groves and leaving behind open, disturbed soils? Adding Big Trees to forests that had not previously supported them also made sense in that it strengthened the species' potential for long-term survival. Substantial segments of the public, however, saw all this in a profoundly different light. What the Forest Service had chosen to ignore was that for nearly a century, mainstream public thought about the sequoias had been shifting ever more toward emphasizing the "sacred" nature of the trees and the corollary that the trees (and the forests in which they grew) ought to be seen as inviolate.

Over time, this perspective had led to the Yosemite Grant, the creation of Sequoia and General Grant National Parks, the public acquisition of the North Grove at Calaveras, and much, much more. It had even played a significant role in the 1893 creation of the Sierra Forest Reserve, which had evolved over time to become the Sequoia National Forest.[14]

The 1980s uproar over government-sponsored logging in giant sequoia groves began within the environmental community and spread quickly. For once, environmental activists had a message that was easy to sell to the larger public. Photographs of towering giant sequoia trees standing alone in acres of clear-cut forest provided rich fodder for those who opposed the program. Such pictures were published in *National Geographic*, *Audubon*, and *Sunset*. In October 1986, under considerable pressure, the supervisor of the Sequoia National Forest suspended further timber sales in all the forest's sequoia groves. The battle, however, was far from over.

Still convinced that it was on the right path, the forest officials moved forward in the late 1980s to finalize a new Land and Resource Management Plan for the Sequoia National Forest. The plan came out in March 1988. Attempting to respond to the storm the agency had provoked, the plan prohibited logging in sequoia groves "except for specific purposes of fuel load reduction, public safety, and maintenance of existing utility easements." Commercial harvesting, the Forest Service made clear, would no longer occur within the groves.[15] What these exclusions actually meant, however, remained uncertain to the agency's critics.

The 1988 forest plan set an annual timber production target for the entire Sequoia National Forest of ninety-seven million board feet and designated 345,000 acres for commercial timber management. These numbers were so large that they had the effect of further enraging the environmental community. Numerous organizations filed appeals objecting to the plan, and pressure on the Forest Service intensified. When the Sierra Club obtained a preliminary injunction blocking portions of the plan, the agency agreed to a mediated settlement negotiation. In July 1990, eighteen parties endorsed the resulting mediated settlement that reduced the forest's timber-management zone by 32 percent and established a new and lower annual production target of seventy-five million board feet. This significant reduction might have been expected to end the controversies surrounding the forest, but it did not. The Forest Service, according to its critics, still wanted to cut trees within its sequoia groves.

By the early 1990s, the Forest Service and the environmental community had settled into the political equivalent of trench warfare. The subject? Giant sequoia forest management. Several environmental organizations took the fight to the public, including the Wilderness Society, the Backcountry Horsemen of America, the California Native Plant Society, and the Sierra Club. On Thursday, September 5, 1991, the Tulare County Audubon Society went so far as to take out a large ad on the subject in the *New York Times*. The lede in the advertisement pulled no punches: "They Were Here Before Christ Was Born, But Will They Survive Smokey the Bear?" That the headline suggested that the Forest Service intended to log even the largest trees was an exaggeration, but it did capture the Audubon Society's sense of impending destruction.

The occasion for this outburst was a congressional oversight hearing the previous day in Visalia focusing on giant sequoia management on the Sequoia National Forest. At the hearing, the Forest Service mounted a vigorous defense of its Big Tree management program, during which it emphasized that the agency had no plans to reverse the 1986 ban on logging mature sequoias but wanted to continue to remove pretty much everything else from the groves. At the hearing, the agency brought in forestry professors from several California universities to argue that so much was yet to be learned about giant sequoias that it would be a mistake to commit to one type of Big Tree management, by which was meant national park–style management that sought to achieve ecological goals through natural processes. Realizing now that public opposition to their management of the forest had grown to be a significant problem, Forest Service officials took the trouble to prepare and distribute a written summary of their testimony at the hearing. The publication included all the agency's position statements together with its rebuttals to the criticisms being made by the environmental community. Not included was testimony made at the hearings by the agency's critics.[16]

Fully engaged now in the defense of its approach to forest management, the Forest Service pursued several very public initiatives. In cooperation with the other agencies involved in giant sequoia management, the

Forest Service took the lead in organizing a major public symposium to discuss the Big Trees and their future. Titled "The Symposium on Giant Sequoias: Their Place in the Ecosystem and Society," the event took place over several days in June 1992 in Visalia. Organized as a major event open to the public, seven multi-speaker panels addressed the full spectrum of giant sequoia issues. Forest Service officials opened and closed the event, but many of the panel presentations included individuals who were strongly critical of the agency's approach to forest management. Reflecting the breadth and diversity of the event, the speakers' list ran to almost fifty names, and several hundred people attended.

Regional Forester Ronald Stewart represented the Forest Service, and he both defended the agency's motives and offered conciliatory words to critics. In describing public reaction to the agency's giant sequoia program in the 1980s, Stewart admitted that "the Forest Service now realizes that it went too far, too fast in implementing management activities in the groves."[17] Throughout its symposium presentations, however, the Forest Service continued to emphasize its belief that the best long-term approach to giant sequoia management would be through forestry and forest manipulation rather than through reliance on the natural-processes approach (e.g., fire) preferred by the National Park Service.

Following closely on the symposium, the Forest Service executed the next major step in its public-relations campaign; it brought in the president of the United States. On July 14, 1992, President George H. W. Bush visited Sequoia National Forest; his goal was to strengthen support for management of the area by the Forest Service. Working with the White House, the Forest Service chose to focus Bush's visit on the Freeman Creek Grove, the largest uncut sequoia grove in the national forest system. Once there (the president flew in by helicopter), Bush signed a presidential proclamation that declared that all the national forest Big Tree groves of the Sierra Nevada "shall be managed, protected, and restored by the Secretary of Agriculture." To clarify the promise being made, the proclamation went on to specify that the groves "shall not be managed for timber production and shall not be included in the land base used to establish the allowable sale quantities for the affected national forests."[18] Once this business was completed, the Forest Service dedicated the "George Bush Tree," a handsome monarch sequoia. The proclamation had the effect of adding presidential weight to the Forest Service's 1986 commitment that it would not cut mature sequoia trees. What no one mentioned that day was that just three years earlier the 1988 version of the Sequoia National Forest Management Plan had called for logging up to 1,580 of the Freeman Grove's 1,800 acres.

If the Forest Service hoped that its symposium and President Bush's proclamation would allow it to regain solid public support, it was disappointed. The political war over the Sequoia National Forest and how to manage its extensive forest resources continued for the rest of the decade. Along the way, the agency endured an evaluation of its management by the congressionally mandated Sierra Nevada Ecosystem Project (which diplomatically concluded that "for the USFS to meet its new mandate, permanent new funding must be earmarked for sequoia management, research and monitoring"[19]), signed an interagency research agreement with the Park Service and several other organizations for giant sequoia studies, and published a book explaining its giant sequoia management goals (*An Ecological Foundation for Management of National Forest Giant Sequoia Ecosystems*).[20] Yet none of this succeeded in quelling environmental concerns about Forest Service management of the sequoias.

Finally, a full twenty-six years after the Forest Service initiated logging in its sequoia groves, a turning point arrived that changed everything for the Sequoia National Forest. On April 15, 2000, President Bill Clinton followed his predecessor onto the embattled forest and issued his own proclamation. Superficially, this event resembled President George H. W. Bush's Big Tree visit eight years earlier, but a closer look quickly showed how much the political situation had changed. In 1992, the Forest Service orchestrated the Bush visit to defend the agency's programs. Eight years later, everything was different. This time, even though Forest Service and Agriculture Department officials stood immediately behind President Clinton as he signed the new proclamation (doing their best to put a positive face on the proceedings), everyone there knew that the president's visit confirmed a profound defeat for the too-proud agency and its ill-fated giant sequoia management program.

Clinton's signing ceremony took place along the Trail of a Hundred Giants, an interpretive trail developed by the Forest Service in the Long Meadow Grove. If anyone remained in doubt, the setting confirmed that this proclamation would be about protecting sequoias for public enjoyment. A look at those attending reinforced the point. Approximately 150 persons were bused to the site, a total that included about 40 from the environmental community.[21] In reality, the day belonged to the environmentalists. For more than a decade they had fought the Forest Service in a long campaign that had been painful for both sides, and they had prevailed.

At the ceremony, President Clinton declared a 327,769-acre Giant Sequoia National Monument that overlaid a new set of rules on a major part of the Sequoia National Forest. Authority for this action came from the Antiquities Act of 1906, a federal statute allowing presidents to give national monument status to federal lands that contain objects of historical and scientific interest. Since the time of Theodore Roosevelt, the act has been used repeatedly to resolve contentious political disputes about whether to protect natural resources. In essence, a presidential decree could be used to make a decision that Congress might not wish to undertake.

The Giant Sequoia National Monument proclamation had some very specific things to say about future management of the area included within its borders. Most importantly, commercial logging would quickly be phased out.

> No portion of the monument shall be considered to be suited for timber production, and no part of the monument shall be used in a calculation or provision of a sustained yield of timber from the Sequoia National Forest. Removal of trees, except for personal use fuel wood, from within the monument area may take place only if clearly needed for ecological restoration and maintenance or public safety.[22]

The scale of what they had just achieved must have amazed many who had fought for it. According to the Forest Service, the Sequoia National Forest contained about 13,200 acres on which giant sequoia trees grew. The new monument included not only those lands but also another 315,000 acres, containing everything from low-altitude oak woodlands to subalpine ridges with thousand-year-old foxtail pines. Until Clinton's proclamation, much of this land had been managed as "capable, available, and suitable" for commercial timber production.[23] Now, or so it appeared on April 15, 2000, that was all to end.

Immediately there arose charges of political motivations. Shaking hands at the rope line after signing the proclamation, Clinton was asked if he was trying to enhance his legacy during the last year of his second term. He defended himself without apparent irony:

> The only issue here is whether we're doing the right thing. I have been doing this kind of thing for seven and a half years now. I've been working on these issues. What I'm trying to do is to build a legacy for these children. And I think we did the right thing.[24]

But of course it was all about politics. Clinton and Vice President Al Gore, who was running to succeed him, were defining themselves in a way they hoped would clearly differentiate them from George W. Bush, the likely Republican nominee for the upcoming presidential election. A gesture to the environmental community seemed appropriate so near to Earth Day, and what could be safer than taking bold action to protect giant sequoia trees, a cause that the nation as a whole had generally agreed on for more than a century? If the locals opposed the proclamation, and many did, there was nonetheless little for Clinton and Gore to lose. Central California had long since defined itself politically as a Republican stronghold.

Surprisingly, however, the fight was still not quite over. When George Bush, and not Al Gore, took office the following January, the balance of power in this endless tree war shifted yet again. Since no legal precedent existed for the presidential de-declaration of a national monument, two other fronts were opened by those who hoped to turn the Sequoia National Forest back toward timber production.

Because of the scale of the lands involved, the fight now was about far more than just the sequoias. Working to overturn the proclamation on legal grounds, the government of Tulare County, the location of most of the monument, filed suit charging that Clinton's decree was too broad and vague to meet the legal requirements of the Antiquities Act, the federal act that gives presidents the authority to create national monuments. Although supported by timber firms and off-road enthusiasts, the suit failed at every level despite repeated appeals of negative decisions. California attorney general Bill Lockyer joined the Sierra Club and the Natural Resources Defense Council in supporting the federal government's position, which was that the declaration met legal standards.

Meanwhile, the Forest Service went to work to implement one of the key clauses in the Clinton proclamation, which was that the agency needed to prepare a management plan for the new monument within three years. Sensing that the Bush administration might allow it to recover some of the lost timber potential of the Sequoia National Forest, the Forest Service began drafting a plan that provided a very generous definition of tree cutting "for ecological restoration and maintenance or public safety." By the time the draft plan came out in 2005, it called for the cutting of 7.5 million board feet of timber annually within the monument, including trees up to thirty inches in diameter. This plan, the Forest Service hoped, would both restore the area's ability to produce timber—still a key national goal for the agency—and meet the requirements set forth in Clinton's proclamation.

But it was not to be. In August 2006, in response to yet another lawsuit, a federal judge ruled the management plan illegal because it failed to carry out the intent of the 2000 presidential proclamation. Sent back to write a new management plan for the monument, the Forest Service tried again. Eventually, after receiving more than 79,000 comments representing every conceivable viewpoint on the plan's draft environmental impact statement, the agency issued a revised plan in September 2012. This new effort promised that all management would occur within the context of twenty-seven advisory points developed by an independent scientific advisory board, and that this time trees really would be removed "only if clearly needed for ecological restoration and maintenance or public safety."[25] The great giant sequoia war, it appeared, had finally ended.

The prolonged disputes over the management of the Giant Sequoia National Monument were ultimately about far more than the 13,000-plus acres of giant sequoias involved. Public acceptance of the "sacred" nature of the Big Trees had grown to be so pervasive that it could be channeled politically at the national level, and the traditional foresters of the Sequoia

National Forest learned this lesson the hard way. In the early 1980s, needing to meet national goals for timber production, they approached the Big Trees groves under their control as if they were forests like any other. By this time, the Forest Service had three-quarters of a century of experience under its belt in presenting its forestry programs to the American public, and it enjoyed a reputation as a savvy and successful federal bureau. It is true that the agency's intense logging program of the 1970s and 1980s had generated controversy and resistance in many regions, but on the Sequoia National Forest issue the agency suffered a spectacular crash, losing not only the public support that makes agency work possible but also eventually management discretion over a huge tract of more than five hundred square miles of forest.

To put it another way, for the Forest Service, the cost of logging about 1,000 acres of giant sequoia forest was the permanent loss of timber production on 327,000 acres. And what did the agency do to deserve such draconian punishment? This can be answered in two ways. The first answer reflects forestry practice. The Forest Service had reasonable evidence to suggest that logging could produce the forest disturbance required to facilitate giant sequoia reproduction. Even the agency's worst critics did not dispute this. Instead these critics focused on the other half of this equation. In its giant sequoia grove logging program, the agency profoundly discounted public sentiments about the Big Trees and the broader settings in which they grew, and it did this not just on the ground between 1980 and 1985 but then again in the 1988 Forest Management Plan, and yet again in the first (2001) management plan for the Giant Sequoia National Monument. In this, the agency displayed a degree of hubris that made its defeat essentially inevitable. Most old-time foresters would have thought it impossible that their agency would be defeated by a tree, but that is essentially what happened in the Sequoia National Forest.

A century and a half after Euro-Americans first encountered the Big Trees at Calaveras, the giant sequoias of the Sierra had grown to hold immense power as a surrogate for the environment as a whole. To large segments of the American public, the trees had become a powerful symbol for nature itself.

EPILOGUE
King Sequoia

> The sequoias do not just make us stop and stare; they make us stop and think, pricking the bubble of our hubris and reminding us of the deeper layers of time all around us. The big trees chant not of the past only, but of the future, like elders pointing the way to wisdom.[1]
>
> —AUTHOR JONATHON SPAULDING

As we move deeper into the twenty-first century, it is hard to deny the catalytic power of the sequoias. In less than two centuries, the trees have risen from simple objects of exploitation to become broadly protected symbols of our relationship with the natural world. During those years, the Big Trees have provoked worldwide curiosity; inspired one of the first American efforts to protect nature for its own sake; been sold wholesale for logging and destruction; shaped the thinking of John Muir (arguably the nation's first modern environmentalist); become the founding focus of our national system of federal parks; taught us the value of ecological thinking; been recognized as natural repositories of information about climate history; provided key perspectives that allowed mainstream American culture to begin to come to grips with ecological processes including fire; and turned the nation's most powerful land management agency and its mission in a new direction. No other native North American tree can claim so influential a role in human affairs.

Yet, it would be a profound mistake to think that this story has reached its conclusion. In many ways, all this is just a beginning. Our time among the giant sequoias has been short, and our knowledge of the trees remains tempered by the relative brevity of our relationship with them. We have set aside most of the natural groves where we first found the trees and done our confused best to manage them. But even as we have protected and treasured the historic Big Trees groves, we humans have set the stage for the new and larger challenges that the trees will have no choice but to face in coming years. The establishment of the national parks and other sequoia preserves of the Sierra Nevada has done little to protect the trees from regional air pollution, which has become severe in the southern portions of the Great Central Valley of California. At the same time, because of human use of the surrounding landscapes, the groves have never been so isolated and at risk; the Sierra's protected landscapes now function as ecological islands. And even these challenges, significant as they are, pale in the face of climate change.

Since the mid-twentieth century, average temperatures in the Sierra Nevada have risen several degrees, and the outlook for continued warming seems inescapable. As temperatures rise, so does the winter snowline in the Sierra and the unpredictability of precipitation of any sort. Increasingly, sequoia groves are seeing winter rain rather than snow, and periods of drought are intensifying. What this may mean for the Big Trees is a question that we have just begun to address, but ecological theory suggests that most of the groves we have worked so hard to protect may be in the wrong places to prosper in the climate in which we will all live in coming centuries.[2] Even as this book goes to print, multiple studies are under way that address the impacts of California's changing climate on the Big Trees. How will the trees fare in the more extreme cycles of flood and drought that seem likely to be the Golden State's future?

Such broad patterns of change will affect all species, of course, but because the Big Trees have such a special relationship with humanity, we must anticipate that our historical concern for them will likely endure and even deepen. Will giant sequoias become symbolic "poster children" in a time of intensifying climate disruption? Will we actively manage the existing old-growth stands to mitigate the destructive effects of a warmer climate by trying to turn such places into "ecosystem museums"? Will we find ourselves seeking out new places for the trees to grow—places where they will have a better chance of surviving in the twenty-second century and beyond?

Only time will tell us if these are even the right questions. Undoubtedly, we have much yet to learn and, just as they have in the past, the sequoias will have much to share with us about both the natural world and ourselves. Since the 1850s, the Big Trees and we humans have become partners in a powerful relationship that continues to grow and evolve. Where that relationship will take us next is far from clear. But what we can know is this: our fascination with these amazing trees and our love for them will lead us into places and ideas we have yet to imagine. So it has always been with the sequoias, and so it will continue to be.

The reign of King Sequoia is far from over.

Acknowledgments

This volume possesses as many supporting roots as does a monarch giant sequoia tree. Without this rich network of support and assistance, the book never could have risen out of the soil in which it first sprouted.

Many of the roots that support a book like this come from the legions of dedicated stewards who work every day to understand, protect, and sustain the Big Trees. At Sequoia and Kings Canyon National Parks, National Park Service employees Ward Eldredge and Tony Caprio were especially helpful to this project. In his role as curator for the parks, Ward Eldredge provided access to and guidance on the rich archival resources of the twin national parks of the southern Sierra. Fire ecologist Tony Caprio reviewed my early efforts to explain how the Big Trees influenced dendrochronology, and he helped focus my thinking in many useful ways.

Finding good illustrations for a historical book is always a challenge, and in addition to the persons already noted above I must give special thanks to Virginia Sanchez at the Yosemite National Park Research Library and Christopher Baisan and Thomas Swetnam at the Laboratory of Tree-Ring Research at the University of Arizona.

Ecologist Nate Stephenson of the Biological Resources Division of the United States Geological Survey is sometimes described by those who work with him as the world's leading expert on the giant sequoias. In this role, which he humbly denies, he provided much useful assistance in everything from identifying research sources to suggesting the illustrations to be included in this volume.

During my research, many others took time to assist me in my work. I would particularly like to thank Joe Fontaine of the Sierra Club and Michael Wurtz of the Holt-Atherton Special Collections at the University of the Pacific, the archive that holds the John Muir Papers.

Gary Lowe, giant sequoia historian par excellence, deserves special recognition here for the amazing work he has done chasing down original materials related to the early decades of this story. Lowe's diligent sleuthing made my work so much easier.

In the growth of every book, there comes a time when it must be read and considered by others. Two early readers of this manuscript, Charles Wollenberg and Jonathon Spaulding, provided feedback that was both productively challenging and highly useful. To both I must offer sincere thanks.

I deeply appreciate the investment made in this book by the Sierra College Press. This began with their commitment to support the publication of this book and progressed through every subsequent phase of the project. Joe Medeiros and Rick Heide deserve special thanks. I am honored to be associated with their publishing program.

And finally, of course, I must thank the amazing staff at Heyday. This book, quite simply, would not exist if Heyday's founder, Malcolm Margolin, had not taken an early interest in it. He played a critical role in defining the book when it was still in the conceptual stage, and he has continued to support it as it has worked its way through the steps through which every book must pass.

Working with the staff at Heyday has been a distinct pleasure. Many, many good people there have invested themselves in the effort to bring this book into the hands of you, the reader. Deserving special recognition are editorial director Gayle Wattawa, art director Diane Lee, and editor Lisa K. Marietta. Each made significant contributions that I deeply appreciate.

Finally, as always, I must thank my ever-so-patient wife and partner, Frances, for enduring all that comes with being the spouse of a would-be writer. She knows, all too well, what I mean.

Notes

CHAPTER 1: The Mammoth Trees of California

1. Zenas Leonard, *Narrative of the Adventures of Zenas Leonard*, as quoted in Farquhar 1965, 37.
2. Stine 2015. The exact date remains mired in myth and confusion. Lowe (2012a, page 23) makes a strong argument for a date soon after a winter storm that swept over the region from December 30 to 31, 1852.
3. Lowe 2012a, 25.
4. Ibid., 31.
5. Engbeck 1988, 73–75. Engbeck also documents that although several other parties had encountered sequoias between 1849 and 1852, their discoveries had been ignored during the early years of the gold rush.
6. Today this name is used as the common name for trees of the genus *Thuja*, which grow in North America and East Asia. The Big Trees did not receive their formal botanical name, *Sequoia gigantea*, until 1854. See Chapter Two for more on this subject.
7. The lithograph is reproduced in Engbeck 1988, 75.
8. Kruska 1985, 18–20.
9. By removing the bark, they killed the tree's cambium layer and thus irreparably damaged its ability to create the new root tissues that conduct minerals and water to the tree's foliage. Such damage to a tree is always ultimately fatal.
10. This and the following paragraphs about Captain Hanford's endeavor all draw upon Lowe 2012a, 39–66, and Kruska 1985, 19–32. Today, more than 150 years later, the grooves left in the wood by the pump augers can still be seen on the butt log of the Discovery Tree.
11. Quoted in Lowe 2012a, 57.
12. Ibid., 76–77.
13. Ibid., 79.

14. Ibid., 94. Here, too, began the enduring habit of exaggerating giant sequoia statistics. The actual tree did not exceed three hundred feet in height, and the described diameter of forty-six feet at two hundred feet is also grossly exaggerated.
15. Ibid., 94.
16. Ibid., 130.
17. Ibid., 113–26 and 146.
18. Ibid., 159–77.
19. Ibid., 316.
20. Kruska 1985, 33–38. Also Lowe 2007, 21–32.
21. Lowe 2007, 31–37.
22. Ibid., 39.
23. Ibid., 40–41.
24. Ibid., 75–90. See also Kruska 1985, 33–38. The Crystal Palace of London had been erected for the pioneering world's fair that occurred in that city in the summer of 1851. It broke architectural ground and generated much excitement for its extensive use of the new material of plate glass. After the fair closed, the structure was purchased by private investors and relocated to suburban Sydenham, where it opened in 1854 as a nineteenth-century amusement park.
25. An 1855 entrance fee of fifty cents is equivalent to approximately fifteen dollars today. A laborer in New York in the 1850s earned on average between one and two dollars per day.
26. This lithograph, issued by Britton and Rey in San Francisco in 1855, is reproduced in Engbeck 1988, page 80.
27. Willis 1859, 385–97.
28. Kruska 1985, 15–17.
29. Willis 1859, 389.
30. Ibid., 389–90.
31. Vischer 1862. Vermaas (2003, pages 57–78) provides a thorough interpretation of Vischer's efforts.
32. These images are explored in considerable detail in Vermaas 2003, 65–67.
33. Again, see Vermaas 2003, 57–78.

CHAPTER 2: To Name Is to Know

1. Beidleman 2006, 361–62. The academy assumed its current name in 1868, when it became the California Academy of Sciences.
2. Vermaas 2003, 11–14. Also Hartesveldt et al. 1975, 22–23.
3. Lowe 2012b, 79, and Lowe 2012c, 9–10.
4. Lowe 2012a, 80–81.
5. Gary Lowe explores this question in detail in Lowe 2012c, 11–12. Steamers left San Francisco twice a month for Central America, where there were connections for Britain. If Lobb sailed from San Francisco on the October 16 "steamer day," he would have disembarked in Falmouth, only one hundred miles from Exeter, on December 3.
6. By the long-established scientific rules of botany, the right to name a new species falls to the first person to describe it formally in a publication.
7. The entire notice is reproduced in Lowe 2012a, 83.
8. The full text of Gray's statement can be found in Lowe 2012a, 312–13. Gray's description was clouded by

the fact that he inspected the imitation sequoia on display in New York in 1854 rather than Captain Hanford's tree.
9. In 1898, George B. Sudworth of the US Department of Agriculture's Division of Forestry made the argument that the tree should be known as *Sequoia washingtoniana*, and this position reigned in official federal documents into the 1920s. See Sudworth 1908, 139. The National Park Service also used *S. washingtoniana* into the late 1920s, but by 1930, federal usage had largely converted to *S. gigantea*. In more recent times, botanists have again separated the giant sequoias from their coastal cousins, and the name currently in favor is *Sequoiadendron giganteum*.
10. Lowe 2012a, 6.
11. These seedlings were apparently grown from seed brought back by Lobb.
12. Lowe 2007, 43–49.
13. Quoted in Vermaas 2003, 15.

CHAPTER 3: A Grove Called Mariposa

1. McFarland 1949, 88.
2. For a more complete accounting of Clark's early years, see Sargent 1964, 15–50.
3. Known as the Mariposa Battalion, this volunteer military group, formed to punish the local Native Americans, had effectively discovered Yosemite Valley in the spring of 1851. For a full account, see Bunnell 1990.
4. Quoted in Sargent 1964, 48–49.
5. Ibid., 52–60.
6. Ibid., 60.
7. Ibid., 61.
8. Ibid., 61. These trees, now known as the Nelder Grove, had seen use by the Southern Miwok people for thousands of years, and it was likely a member of that Native American group who guided Clark to the site. Other Euro-Americans had also visited the site before Clark, including members of the Mariposa Battalion in 1851. (See note 3 above.)
9. The Fresno Grove was then in Fresno County. Today it is known as the Nelder Grove, named after a resident of the area during the nineteenth century.
10. Quoted in Sargent 1964, 61.
11. Ogden 1993, 50–51. Photography historian Weston Naef has speculated that these photographs were actually taken by Carleton Watkins.
12. Naef 2008, 4–5 and plates 1 and 2.
13. Ogden 1993, 51.
14. This and the following quote come from Greeley 1860, Chapter 29. We must assume that Greeley did not find the damaged Discovery Tree or the Mother of the Forest to be among the specimens he reported as being in such good condition at Calaveras.
15. King 1962, 31. King's letters to the *Boston Evening Transcript* originally ran between December 1, 1860, and February 9, 1861. A century later, in 1962, the letters were published in book form under the title *A Vacation among the Sierras: Yosemite in 1860*.

16. Quoted in Huth 1948, 67. Many accounts discuss the creation of Yosemite as a public park; in addition to Huth's pioneering statement, see Runte 1990.
17. Jones 1965, 28–29.
18. Olmsted 2009, 6.
19. This and the following quotes on this subject are from Runte 1990, 20.

Chapter 4: An Arboreal Mecca

1. Sargent 1979, 39.
2. *An Act Authorizing a Grant to the State of California of the "Yo-Semite Valley" and of the Land Embracing the "Mariposa Big Tree Grove,"* 13 Stat. 325, Section 1.
3. Sargent 1964, 67–68.
4. Ibid., 70.
5. Ibid., 90–91.
6. Ibid., 121–22. Clark would be reappointed Yosemite's guardian in 1889 and go on to serve a second seven-year term in that position.
7. Sargent 1979, 33.
8. Ibid., 39. Sargent quotes a source who suggests that the word actually came from the call of the owls that lived within the grove.
9. Hutchings 1862, Chapter 5.
10. Whitney 1869. Although the luxurious first printing was quite small (only 250 copies), the text was revised two years later into a more compact guidebook form, which itself went through several subsequent editions and became a defining document for the Sierra Nevada.
11. He also provided a surprisingly accurate statement about the existence of large numbers of sequoias in the watersheds immediately to the south of the main fork of the Kings River.
12. Vermaas 2003, 54–55.
13. For the 1861 photo, see Naef 2008, plate 2. The 1866 image is reproduced in Sargent 1964, 54.
14. His 1863 painting *The Rocky Mountains, Lander's Peak* sold two years later for $25,000, a sum equal to about $370,000 in 2015.
15. Vermaas 2003, 85.
16. The painting is privately held and not available for viewing, but electronic images of it can be found easily on the web.
17. Officially, the "International Exhibition of Arts, Manufactures and Products of the Soil and Mine," this was the country's first world's fair.
18. Quoted in Vermaas 2003, 88.

Chapter 5: Yet Grander Forests

1. In 1852, Mariposa County was divided and the southern portion became Tulare County, a new legal entity named after *los tulares*, the Spanish-language name for the extensive swamplands found there.
2. Grey, n.d., 8. Today this site is known as Grant Grove.
3. Fry 1924, 1.
4. Otter 1963, 22.
5. Whitney 1869, Chapter 5.
6. Lowe 2004, 13–14.
7. Quoted in Lowe 2004, 14.

8. Quoted in Lowe 2004, 120.
9. Lowe 2004, 119–20.
10. Lowe 2004, 120. Girdling a tree involves cutting through the bark and into the sapwood beneath in a way that prevents the tree from using the sapwood to move fluids between the roots and the foliage.
11. Ibid., 19–21.
12. Lowe (2004, pages 26–109) lays out the entire sequence of events in great detail.
13. Ibid., 112–14.
14. Quoted in Lowe 2006, 14.
15. Quoted in Lowe 2006, 15. The letter ran in the *Daily Evening Bulletin* for August 13, 1875.

CHAPTER 6: A Wandering Scot

1. Engberg 1984, 137.
2. Muir 1878, 821.
3. Sargent 1964, 81.
4. Badé 1923, Volume 1, 270–71.
5. Gisel and Joseph 2008, 104*ff*.
6. Muir's letter about "sequoia wine," quoted above, was addressed to Mrs. Carr. In 1875, Jeanne Carr would introduce Muir to Louisa Strenzel, the women he would marry in April 1880.
7. Worster 2008, 205.
8. Gray 1872, 578.
9. Ibid., 582. In this question, Gray anticipated the results of the twentieth-century paleontological work that found fossils of sequoia ancestors scattered geographically across an arc from Wyoming to California.
10. Ibid., 579.
11. The *Glyptostrobus pensilis*, known commonly as the Chinese swamp cypress, is found naturally in southeastern China and extreme northern Vietnam. The genus was named by Steven Endlicher of "sequoia" fame.
12. Ibid., 594–96. This conclusion remains essentially accepted today.
13. *The American Naturalist*, first published in 1867, continues to be issued as a scholarly journal by the American Society of Naturalists and the University of Chicago Press.
14. Kimes and Kimes 1986, number 2.
15. Ibid., numbers 24, 26, 28, 29, 32, and 33.
16. Ibid., number 34.
17. Those wishing to trace Muir's endless literary reworking of this material can start with Engberg 1984, which reprints the articles that ran in late 1875 in the *Daily Evening Bulletin* (see pages 123–37). A mature version of the same materials can be found in Muir's 1901 book, *Our National Parks* (see Chapter 9), which is itself based on essays Muir prepared for the *Atlantic Monthly* after 1900.
18. Engberg 1984, 123–29.
19. This grove, which Muir called "Dinky Grove" because it was located along a creek bearing that name, is now known as the McKinley Grove. (Today called Dinkey Creek, the name gained its extra "e" after Muir's visit.)
20. Mill Flat now forms the floor of the small reservoir known as Sequoia Lake.
21. Muir 1901, 297–98. The site was later lost, and it was not relocated

until the 1970s. The dead tree is now known unofficially as the Muir Snag. Muir noted in his writings that he had found it difficult to get an exact ring count of the tree because there were several points at which it was difficult to interpret zones that showed regrowth following fire damage.
22. This grove had earlier been known as the Tulare Grove and is now the Grant Grove of Kings Canyon National Park.
23. Muir 1878, 821.
24. Muir 1901, 299. Hyde's Mill operated until 1879, and shortly after its closure a Tulare County resident named Horace Whitaker acquired the lands in question. Whitaker allowed no further logging and, upon his death in 1910, the land was donated to the University of California with deed restrictions requiring that no additional sequoias be cut, that the university use the 320-acre tract for forestry investigations, and that the site be known as "Whitaker's Forest." Whitaker's Forest continues to be managed today by the University of California under those conditions.
25. Muir's record here is confusing. In both his newspaper articles of late 1875 and the several longer essays he wrote soon thereafter, he writes about "the giant forest" (without capital letters), apparently referring to the extensive forests of sequoias south of the Kings River, as well as "The Giant Forest" (capitalized), a specific grove located between the Marble and Middle Forks of the Kaweah River. Not until 1901 would Muir publish his claim to have named the Giant Forest of Sequoia National Park (Muir 1901, 300).
26. Engberg 1984, 134 and 136.
27. Muir's essay, titled "On the Post-Glacial History of Sequoia Gigantea," would be published the following summer in the latest issue of the *Proceedings of the American Association for the Advancement of Science*. See Muir 1877.
28. Muir 1877, 3.
29. Kimes and Kimes 1986, number 52.
30. Muir 1878.
31. Muir 1876.
32. Grunsky 1968, 244.

CHAPTER 7: Free for the Taking

1. Willard 1995, 63.
2. Wolfe 1938, 229.
3. Willard 1995, 138–39.
4. Otter 1963, 47–48.
5. For a more complete exploration of Timber and Stone Act fraud, see Ise 1920, 70–83.
6. Created during the presidency of Thomas Jefferson, the township-and-range system surveyed the public domain into "townships" containing thirty-six square miles each.
7. Otter 1963, 68–69.
8. Ibid., 96.
9. Muir 1901, 300.
10. Weiner 2009, 37–53.
11. Willard 1995, 90–91.
12. Johnston 1966, 23–25.
13. Their investment of $1,000,000 in 1888 would equal roughly $25,000,000 in 2015.

14. Johnston 1966, 32.
15. Ibid., 49–50.
16. This extension took the form of an incline railway that climbed over what became known as Hoist Ridge.
17. Ibid., 59.
18. Willard 1995, 78; Johnston 1966, 81.
19. This attitude and the images it produced are captured especially well in Johnston 1966 and Weiner 2009.
20. Fry and White 1930, 20.
21. The most intensive sequoia logging during this second wave of activity took place in the Evans Grove from 1914 to 1917. According to Willard (1995, pages 113–20), only the logging in Converse Basin exceeded what was done to the sequoias in Evans Grove.
22. O'Connell 1999, 30–31.
23. Ibid., 34–35.
24. Ibid., 34–35.
25. Ibid., 41.

CHAPTER 8: Of Tunnel Trees and National Parks

1. Historically, the primary significance of the grove had been the possibility that the Big Trees of the Tuolumne Grove were those encountered in 1833 by the Walker Party during their crossing of the Sierra, an event that produced the first written mention of the giant sequoias. This interpretation has now been effectively demolished by Scott Stine in his careful analysis of Walker's route. See Stine 2015.
2. This photo can now be seen as part of the National Park Service exhibit at the site.
3. As was the case with the Dead Giant of the Tuolumne Grove, the passage through the Wawona Tunnel Tree was made by enlarging an existing fire scar. The opening would not prove sustainable over the long run, however, and the tree ultimately collapsed in February 1969 while heavily weighted with snow. The cause of the collapse was a structural failure that began along the upper edge of the tunnel cut.
4. Unlike its more famous cousin, this other tunnel tree at Wawona still stands. It is known as the California Tunnel Tree.
5. They did, however, share the same source of inspiration: Yellowstone National Park. It had been created out of public land by Congress in 1872 and was retained under federal control as a "national" park simply because there was as yet no state of Wyoming to manage the land. It is interesting to speculate whether the national park idea would ever have developed as it did in the United States had there been a state of Wyoming in 1872. If this had been the case, Yellowstone would likely have been given over to state control, as happened to California's Yosemite in 1864.
6. Strong 1964, 66–67.
7. Around this same time, another, much more ambitious effort to protect the southern Sierra was less successful. In December 1881, California senator John Miller proposed the creation of a huge national park covering sixty-two townships of land. The area would cover large portions of the Kings, Kaweah, Tule,

and Kern Rivers' watersheds and would have protected the heart of the giant sequoia range from the provisions of the Timber and Stone Act. Unfortunately, the bill died in committee, essentially ignored by all but a few local backers (Strong 1964, 70).
8. Strong 1964, 84–86.
9. Dilsaver and Tweed 1990, 61–69.
10. Jones 1965, 42–43.
11. Ibid., 44–45; *An Act to Set Apart Certain Tracts of Land in the State of California as Forest Reservations*, 26 Stat. 650.
12. Berland 1962, 79.
13. Orsi 2005, 365.
14. Dilsaver and Tweed 1990, 73.
15. Berland 1962, 79.
16. Currey and Kruska 1992, items 305–21.
17. Dilsaver and Tweed 1990, 75.
18. Sherwood 1927, 4.
19. Weiner 2009, 97.
20. Kruska 1985, 53.
21. The fair had originally been scheduled for the summer of 1892 but was delayed a year when preparations fell behind schedule.
22. Weiner 2009, 102–19. Excellent photos taken by Curtis document every step of this work.
23. Vermaas 2003, 129. The two cross sections removed from the Mark Twain Tree remain on exhibit today at the American Museum of Natural History in New York City and the Museum of Natural History in London. The stump of the Mark Twain Tree now falls within Kings Canyon National Park and is a historical feature along the trail that explores Big Stump Basin. After the closure of the World's Columbian Exposition, the Interior Department gave the exhibit they had created from the General Noble Tree to the Smithsonian Institution, which displayed it on the National Mall in Washington, DC, until that area was covered with temporary offices during the Second World War. The exhibit went into storage at that time and disappeared. All that remains today of the General Noble Tree is its cut base, now known as the "Chicago Stump" and located in the Converse Basin area of the Sequoia National Forest.

CHAPTER 9: For the Greater Good

1. Quoted in Grunsky 1968, 245.
2. Muir 1912, 139.
3. Runte 1990, 59–64.
4. Strong 1964, 140.
5. Kimes and Kimes 1986, number 185.
6. Strong 1964, 142–44.
7. Ibid., 144.
8. Significant portions of the sequoia groves within the new reserve were still in government hands when the reserve was created, meaning their withdrawal from sale was all the more timely. Later, many of the tracts that had been sold before 1893 would drift back into public ownership, some after they had been logged, but others with at least some of their sequoias still standing. (See Chapter 14.)
9. US House of Representatives 1900, 8.
10. US House of Representatives 1900, 2.

11. Engbeck 1988, 95–96.
12. US Department of Agriculture 1900, maps inserted following page 22.
13. Engbeck 1988, 96–98.
14. Jones 1965, 55–63.
15. Johnston 2008, 231–33.
16. Ibid., 233–36.
17. Jones 1965, 63–79.

CHAPTER 10: A Source of Inspiration

1. Currey and Kruska 1992, 150.
2. Wells 1907, inside back cover.
3. Southern Pacific Company 1914.
4. Wells 1907, 31.
5. Southern Pacific Company 1914, 11.
6. Southern Pacific Company 1914, 18.
7. Santa Fe Railway 1901, 48–49.
8. Clark 1907.
9. Clark 1904.
10. Clark 1907, 90–91.
11. Ibid., 104.
12. Sargent 1964, 154 and 164–66.
13. Clark 1907, 15.
14. One of Muir's oft-noted quotes expounds: "Writing is like the life of a glacier; one eternal grind."
15. Kimes and Kimes 1986, number 189.
16. Modern estimates give the trees a life span of about 3,200 years.
17. Muir 1894, 179–203.
18. Kimes and Kimes 1986, numbers 234 and 237.
19. Muir 1901, 268.
20. In his writings, Muir never identified his host. Tharp owned the meadow where the encounter occurred, but Wolverton managed the site and was there much more than his boss.
21. Muir 1901, 300.
22. Ibid., 298 and 307.
23. Ibid., 300–301.
24. Ibid., 302–3.
25. Kimes and Kimes 1986, number 308.
26. For a short biography and appreciation of Chase, see Powell 1989, 197–207.
27. Chase 1911.
28. Ibid., 136–37.
29. Ibid., 140-41.
30. Driesback 1997, 75–84.
31. Vermaas 2003, 135–37.
32. Driesback 1997, 120.
33. Albright and Shenck 1999, 30–37.
34. Albright and Shenck 1990, 5–6.
35. Albright and Shenck 1999, 69.

CHAPTER 11: Science and Time

1. Johnston 1966, 97–124.
2. Willard 1995, 115.
3. Huntington 1907.
4. McGraw 2003, 23.
5. Douglass 1909.
6. Clark 1907, 53.
7. Huntington 1913, 12.
8. Ibid., 12–18.
9. Ibid., 20.
10. As noted elsewhere, Huntington's ring counts were more indicative of sequoia ages as they are now understood than he wanted to believe at the time.
11. Huntington 1913, 24.
12. Huntington 1914.
13. Douglass 1919, 47.
14. Ibid., 48, figure 12.

15. This represented the remains of the Daniel Webster Tree that had been cut by Martin Vivian in 1875.
16. Douglass 1919, 52, table 5.
17. Douglass 1928, 51–54.

CHAPTER 12: Running into Limits

1. Sargent 1975, 49–63.
2. Greene 1987, 359.
3. This was the beginning of Yosemite National Park's famous High Sierra Camp network.
4. A more complete history of early concessions development in Sequoia National Park can be found in Tweed 1972.
5. Dilsaver and Tweed 1990, 112.
6. Meinecke 1926, 1.
7. Ibid., 2.
8. Ibid., *passim*.
9. Ibid., 11–12.
10. National Park Service 2004, Part 2a, 12.
11. Meinecke 1927.
12. Hartesveldt 1962, 52.
13. Ibid., 54–56.
14. Quoted in Hartesveldt 1962, 193.
15. Hartesveldt 1962, 189–99.
16. Dilsaver and Tweed 1990, 149.
17. Ibid., 151–52.
18. Welcomed at the time as a necessary improvement, this water system ultimately would have significant unanticipated consequences when its overflow drain created an artificial wetland that threatened a part of the forest. See Chapter 16.
19. National Park Service 2004, Part 2a, 23.
20. Good 1938, Part 3, 70–71.
21. Hays 1934.

CHAPTER 13: Words as Grand as Trees

1. US Department of the Interior 1915, 7–8.
2. US Department of the Interior 1921, 16–21.
3. Huntington 1913.
4. Hall 1920.
5. Hall 1921a.
6. Hall 1921b.
7. This pattern would continue through at least 1924. See US Department of the Interior 1924, 10–14.
8. Willis Linn Jepson (1867–1946) taught botany at the University of California from 1899 until 1937 and is generally recognized as California's most significant early-twentieth-century botanist. The Jepson Herbarium at the University of California, Berkeley, is named in his honor, as is the *Jepson Manual*, the contemporary reigning botanical reference guide to California's native flora.
9. We'll explore this in more detail in Chapter 15.
10. Hall 1921b, 235–46.
11. Ellsworth 1924.
12. Ibid., 13.
13. Ibid., 101.
14. Ibid., 106.
15. Ibid., 133.
16. Wilson 1922.
17. Wilson 1928.
18. Ibid., 51.
19. Ibid., 52.
20. Ibid., 73.

21. Ibid., 88.
22. Ibid., 89.
23. Wilson's 1920s books on both Sequoia and Yosemite National Parks are today perhaps the easiest of all similar publications of this era to find on the used-book market.
24. Stewart 1930.
25. Ibid., 95.
26. Fry and White 1930.
27. Dilsaver and Tweed 1990, 101–3 and 111–12.
28. Fry and White 1930, 6–7.
29. Ibid., 37–38.
30. Ibid., 35–36.
31. Ibid., 62–63.
32. Ibid., 92–93.
33. Jepson 1923, 23.
34. Fry and White 1930, 97.
35. Ibid., 100–108.
36. Today, most students of the Big Trees recognize between sixty-five and seventy groves. See Willard 1995 and 2000.
37. Fry and White's *Big Trees* remained on sale in national park visitor centers for the next forty years. A purchase date written on the inside cover of this author's first copy (1938 revision, seventh printing) shows that it was purchased in Sequoia National Park in June 1969.

CHAPTER 14: Belonging to All

1. Dilsaver and Tweed 1990, 113.
2. *National Geographic Magazine* 1916.
3. Dilsaver and Tweed 1990, 113. The National Geographic Society continued to support land acquisition in Sequoia National Park. By the early 1920s, the society had purchased and donated to the park 1,916 acres at a cost of $96,330. A plaque commemorating the 1916 donation can still be found along the western edge of the grove's Round Meadow.
4. Ibid., 113–18.
5. Ibid., 116. During various reorganizations, the Hume-Bennett Lumber Company had reverted to its original name.
6. Farquhar 1941, 48–49.
7. Johnston 1966, 134–135.
8. Otter 1963, 118–122.
9. Engbeck 1988, 98–100.
10. Organized in 1918 to protect threatened groves of coast redwoods, the Save the Redwoods League soon extended its mission to protect the giant sequoias as well. During this same period, Rockefeller purchased more than 35,000 acres of land for the purpose of adding it to Grand Teton National Park in Wyoming.
11. Engbeck 1988, 100.
12. Sargent 1979, 67–69.
13. California State Park Commissioner 1952, 31.
14. US Department of Agriculture 1900, 25–26.
15. Southern Pacific Company 1914, 5.
16. Van Name 1927, 3–5.
17. Dilsaver and Tweed 1990, 209.
18. California State Park Commissioner 1952, 29.
19. US Congress, Committee on the Public Lands 1939, 171.
20. Ibid., 9.
21. Otter 1963, 128–30.
22. Ibid., 131–33.
23. Ibid., 134–39.
24. Ibid., 135.
25. Otter and Dulitz 2007, 25–28.
26. Engbeck 1988, 102.

27. A more complete explanation of this complicated campaign can be found in Engbeck 1988, 101–10.
28. California State Park Commissioner 1952, vi.
29. Otter 1963, 124.
30. Engbeck 1988, 107.
31. Brown 1951, 679; Engbeck 1988, 107.
32. US Department of Agriculture 1900.
33. California State Park Commissioner 1952, 10.
34. Ibid., 25.
35. Ibid., 67. Despite the recommendations of the state report, the Black Mountain and Freeman Creek Groves never became state parks. Instead they remained part of the Sequoia National Forest, which worked over succeeding years to extinguish the private holdings within the groves. In 2011, the Save the Redwoods League purchased yet another sixty-acre inholding within the Black Mountain Grove and donated it to the US Forest Service.
36. Engbeck 1988, 109–11.
37. California State Park Commissioner 1952, 5.

CHAPTER 15: Kindled Light

1. All quotes in this paragraph from Muir 1894, 191.
2. Sudworth 1908, 145.
3. Ibid., 145.
4. Pinchot 1899, 77.
5. Pinchot 1907, 18.
6. For the full story of this great fire and its significance, see Egan 2009.
7. Hall 1921b, 243.
8. Ellsworth 1924, 99.
9. Fry and White 1930, 60.
10. Ibid., 85.
11. National Park Service 2004, Part 2b, 1.
12. As late as 1957, Park Service biologists were still assigned to a branch within the agency's interpretive (naturalist) function. Park naturalists were hired for their biological qualifications and were expected to be capable of addressing ecological questions on the occasions that such things were considered.
13. Hartesveldt 1962, 62.
14. Ibid., 2.
15. Ibid., vii and 2.
16. Ibid., 65–121.
17. Ibid., 122–24.
18. Ibid., 128–31.
19. Ibid., 139–49.
20. All together, Hartesveldt identified thirteen recommendations (1962, pages 214–32).
21. Haertesveldt 1962, 231.
22. Today, this institution is known as San Jose State University.
23. Even before Hartesveldt began his formal studies in Kings Canyon in 1964, he had been paid by the Sequoia Natural History Association to make an analysis of visitor impacts within the Giant Forest. See Harvey et al. 1980, xvii.
24. A full account of this study can be found in Harvey et al. 1980.
25. Biswell 1959.
26. Biswell 1961.
27. In 1875, when Muir visited this site, Hyde's Mill was cutting sequoias in this part of the Redwood Mountain Grove. The tract was donated to the University of California in 1910.

28. Carle 2002, 121–23 and 127.
29. A. Starker Leopold was the son of pioneering ecologist and bioethicist Aldo Leopold, author of *A Sand County Almanac*.
30. Sellars 1997, 214–16.
31. Ibid., 253–54.
32. Carle 2002, 133–35.
33. Kilgore 1970.
34. Hartesveldt et al. 1975.
35. Harvey et al. 1980.
36. Harvey et al. 1981.

CHAPTER 16: Worth the Fight

1. Willard 1995, 263–65.
2. Ibid., 264.
3. National Park Service 2004, Part 2b, 5.
4. An intense two-day storm in late February 1969 deposited up to twelve feet of snow in the Sierra Nevada, one of the heaviest snowfalls ever recorded on the planet.
5. National Park Service 2004, Part 2b, 5. The lodge building endured for another decade as a dormitory for the Youth Conservation Corps. After being damaged by fire, it was finally removed in about 1983.
6. Dilsaver and Tweed 1990, 244–46.
7. Ibid., 289–95.
8. For the record, the author was deeply involved in this effort, heading the park's planning office from 1988 until 1996.
9. California State Park Commissioner 1952, 35.
10. In 1970, President Richard Nixon ordered through executive action that national-forest timber sales be increased by 50 percent. This initiative, drafted by the forest products industry, was called the Forest and Related Resources (FARR) Plan. The pressure on Forest Service officials during this time to "get out the cut" was substantial.
11. US Forest Service 1991, "Briefing Paper for Congressional Hearing on Giant Sequoia, Sequoia National Forest," 4.
12. Ibid., 5.
13. Ibid., 2.
14. For a more complete statement on this point, see Tweed 1992.
15. US Forest Service 1991, "Giant Sequoia Questions for Sequoia National Forest," 1.
16. US Forest Service 1991, "Giant Sequoia Questions for Sequoia National Forest."
17. US Forest Service 1994, 155.
18. Bush 1992.
19. Stephenson 1996, 1461.
20. Piirto and Rogers 1999.
21. Wood 2000.
22. Clinton 2000.
23. "Response to Testimony of Carla Cloer," in US Forest Service 1991.
24. White House 2000.
25. US Forest Service 2012.

EPILOGUE

1. Personal correspondence, April 29, 2015.
2. For a fuller exploration of these questions and their implications, see this author's *Uncertain Path: A Search for the Future of National Parks* (Tweed, 2010).

Bibliography

Albright, Horace M., and Marian Albright Shenck. 1990. *The Mather Mountain Party of 1915*. Three Rivers, CA: Sequoia Natural History Association.

———. 1999. *Creating the National Park Service: The Missing Years*. Norman: University of Oklahoma Press.

Badé, William. 1923. *The Life and Letters of John Muir: In Two Volumes*. Boston: Houghton Mifflin.

Beidleman, Richard G. 2006. *California's Frontier Naturalists*. Berkeley: University of California Press.

Berland, Oscar. 1962. "Giant Forest's Reservation: The Legend and the Mystery," *Sierra Club Bulletin* 47 (9): 68–82.

Biswell, Harold H. 1959. "Man and Fire in Ponderosa Pine in the Sierra Nevada of California," *Sierra Club Bulletin* 44 (7): 44–52.

———. 1961. "Big Trees and Fire," *National Parks Magazine* 35 (April): 11–14.

Brown, Andrew. 1951. "Saving the Earth's Oldest Living Things," *National Geographic Magazine* 99 (May): 679–95.

Bunnell, Lafayette H. 1990. *The Discovery of Yosemite and the Indian War of 1851 Which Led to that Event*. Yosemite National Park: Yosemite Association. [This edition is based on the 1911 "fourth edition."]

Bush, George H. W. 1992. Proclamation. "Giant Sequoia in National Forests, Proclamation 6457." July 14. [Online at http://www.presidency.ucsb.edu/proclamations.php?year=1992.]

California State Park Commissioner and State Forester. 1952. "The Status of *Sequoia Gigantea* in the Sierra Nevada." Mimeographed report to the California state legislature. Sacramento: California Department of Natural Resources.

Carle, David. 2002. *Burning Questions: America's Fight with Nature's Fire.* Westport, CT: Praeger.

Chase, J. Smeaton. 1911. *Yosemite Trails: Camp and Pack-Train in the Yosemite Region of the Sierra Nevada.* Boston: Houghton Mifflin.

Clark, Galen. 1904. *Indians of Yosemite Valley and Vicinity: Their History, Customs and Traditions.* San Francisco: H. S. Crocker.

———. 1907. *The Big Trees of California: Their History and Characteristics.* Redondo, CA: Reflex Publishing Company.

Clinton, William J. (Bill). 2000. Proclamation. "Establishment of the Giant Sequoia National Monument by the President of the United States, Proclamation 7295." April 15. [Online at http://www.presidency.ucsb.edu/ws/?pid=62321.]

Currey, Lloyd, and Dennis Kruska. 1992. *Bibliography of Yosemite, the Central and Southern High Sierra, and the Big Trees: 1839–1900.* Los Angeles: Dawson's Book Shop.

Dilsaver, Lary M., and William C. Tweed. 1990. *Challenge of the Big Trees: A Resource History of Sequoia and Kings Canyon National Parks.* Three Rivers, CA: Sequoia Natural History Association.

Douglass, Andrew Ellicott. 1909. "Weather Cycles in the Growth of Big Trees," *Monthly Weather Review* 37: 226–37.

———. 1919 and 1928. *Climatic Cycles and Tree-Growth: A Study of the Annual Rings of Trees in Relation to Climate and Solar Activity.* Two volumes. Washington, DC: Carnegie Institution.

Driesback, Janice. 1997. *Direct from Nature: The Oil Sketches of Thomas Hill.* With an essay by William Gerdts. Yosemite National Park and Sacramento: Yosemite Association in association with the Crocker Art Museum.

Egan, Timothy. 2009. *The Big Burn: Teddy Roosevelt and the Fire That Saved America.* Boston: Houghton Mifflin Harcourt.

Ellsworth, Rodney Sydes. 1924. *The Giant Sequoia: An Account of the History and Characteristics of the Big Trees of California.* Oakland, CA: J. D. Berger.

Engbeck, Joseph H., Jr. 1988. *The Enduring Giants: The Epic Story of Giant Sequoia and the Big Trees of Calaveras.* Third edition. Sacramento: California Department of Parks and Recreation.

Engberg, Robert, ed. 1984. *John Muir: Summering in the Sierra.* Madison: University of Wisconsin Press.

Farquhar, Francis. 1941. "Legislative History of Sequoia and Kings Canyon National Parks," *Sierra Club Bulletin* 26 (1): 42–58.

———. 1965. *History of the Sierra Nevada.* Berkeley: University of California Press.

Flint, Wendell. 2002. *To Find the Biggest Tree.* Three Rivers, CA: Sequoia Natural History Association.

Fry, Walter. 1924. "The Discovery of Sequoia National Park and the Sequoia Groves of Big Trees It Contains," *Sequoia Nature Guide Service, Bulletin Number 1,* November 22.

Fry, Walter, and John R. White. 1930. *Big Trees.* Stanford: Stanford University Press.

Gisel, Bonnie J., and Stephen J. Joseph. 2008. *Nature's Beloved Son: Rediscovering John Muir's Botanical Legacy.* Berkeley: Heyday.

Good, Albert. 1938. *Park and Recreation Structures.* Three volumes. Washington, DC: National Park Service.

Gray, Asa. 1872. "Sequoia and Its History," *American Naturalist* 6 (10).

Greeley, Horace. 1860. *An Overland Journey from New York to San Francisco in the Summer of 1859.* New York: C. M. Saxton, Barker, and Co. [Online at http://www.yosemite.ca.us/library/greeley.]

Greene, Linda W. 1987. *Yosemite: The Park and Its Resources.* Denver: National Park Service.

Grey, Fern. n.d. *And the Giants Were Named.* Three Rivers, CA: Sequoia Natural History Association.

Grunsky, Frederick. 1968. *South of Yosemite: Selected Writings by John Muir.* Garden City, NY: Natural History Press.

Hall, Ansel. 1920. *Guide to Yosemite: A Handbook of the Trails and Roads of Yosemite Valley and the Adjacent Region.* San Francisco: Sunset Publishing House.

———. 1921a. *Guide to the Giant Forest, Sequoia National Park: A Handbook of the Northern Section of Sequoia National Park and the Adjacent Sierra Nevada.* Yosemite National Park: self-published.

———, ed. 1921b. *Handbook of Yosemite National Park: A Compendium of Articles on the Yosemite Region by the Leading Scientific Authorities.* New York: G. P. Putnam's Sons.

Hartesveldt, Richard J. 1962. "The Effects of Human Impact upon *Sequoia Gigantea* and Its Environment in the Mariposa Grove, Yosemite National Park, California." PhD diss., University of Michigan.

Hartesveldt, Richard J., H. Thomas Harvey, Howard S. Shellhammer, and Ronald E. Stecker. 1975. *The Giant Sequoia of the Sierra Nevada.* Washington, DC: National Park Service.

Harvey, H. Thomas, Howard S. Shellhammer, and Ronald E. Stecker. 1980. *Giant Sequoia Ecology: Fire and Reproduction.* Washington, DC: National Park Service.

Harvey, H. Thomas, Howard S. Shellhammer, Ronald E. Stecker, and Richard J. Hartesveldt. 1981. *Giant Sequoias.* Three Rivers, CA: Sequoia Natural History Association.

Hays, Howard. 1934. Letter to A. B. Cammerer, August 9, 1934. In the museum collection at Sequoia and Kings Canyon National Parks.

Huntington, Ellsworth. 1907. *The Pulse of Asia: A Journey in Central Asia Illustrating the Geographic Basis of History.* Boston: Houghton Mifflin.

———. 1913. *The Secret of the Big Trees.* Washington, DC: GPO. [Originally published under the same title in *Harper's Magazine* in July 1912.]

———. 1914. *The Climatic Factor as Illustrated in Arid America.* Washington, DC: Carnegie Institution.

Hutchings, James M. 1862. *Scenes of Wonder and Curiosity in California.* San Francisco: J. M. Hutchings. [Online at http://www.yosemite.ca.us/library/scenes_of_wonder_and_curiosity.]

Huth, Hans. 1948. "Yosemite: The Story of an Idea," *Sierra Club Bulletin* 33 (March): 47–78.

Ise, John. 1920. *The United States Forest Policy.* New Haven, CT: Yale University Press.

Jepson, Willis. 1923. *The Trees of California.* Berkeley: Sather Gate Bookshop.

Johnston, Hank. 1966. *They Felled the Redwoods: A Saga of Flumes and Rails in the High Sierra.* Los Angeles: Trans-Anglo Books.

———. 2008. *The Yosemite Grant, 1864–1906: A Pictorial History.* Yosemite National Park: Yosemite Association.

Jones, Holway R. 1965. *John Muir and the Sierra Club: The Battle for Yosemite*. San Francisco: Sierra Club.

Kilgore, Bruce. 1970. "Restoring Fire to the Sequoias," *National Parks Magazine* 44 (277): 16–22.

Kimes, William, and Maymie Kimes. 1986. *John Muir: A Reading Bibliography*. Fresno, CA: Panorama West Books.

King, Thomas Starr. 1962. *A Vacation among the Sierras: Yosemite in 1860.* Edited and with an introduction by John Adams Hussey. San Francisco: Book Club of California. [Online at http://www.yosemite.ca.us/library/vacation_among_the_sierras.]

Kruska, Dennis G. 1985. *Sierra Nevada Big Trees: History of Exhibits, 1850–1903.* Los Angeles: Dawson's Book Shop.

Lowe, Gary D. 2004. *The Big Tree Exhibits of 1870–1871 and the Roots of the Giant Sequoia Preservation Movement.* Second edition. Livermore, CA: Lowebros Publishing.

———. 2006. *The Grant Grove Centennial Stump, Kings Canyon National Park, and the Big Tree Exhibit of 1875–1876.* Livermore, CA: Lowebros Publishing.

———. 2007. *The Mammoth Tree: A Gold Rush Adventure; History of the Mother of the Forest Exhibit of 1854–1866.* Livermore, CA: Lowebros Publishing.

———. 2012a. *The Original Big Tree.* Livermore, CA: Lowebros Publishing.

———. 2012b. *The Discoveries of the Giant Sequoia.* Livermore, CA: Lowebros Publishing.

———. 2012c. *William Lobb: Giant Sequoia Expressman.* Livermore, CA: Lowebros Publishing.

———. 2012d. "Endlicher's Sequence: The Naming of the Genus Sequoia," *Fremontia* 40 (1–2): 25–35.

McFarland, James W. 1949. "A Guide to the Giant Sequoias of Yosemite National Park," *Yosemite Nature Notes* 28 (6).

McGraw, Donald. 2003. "Andrew Elliott Douglass and the Giant Sequoias in the Founding of Dendrochronology," *Tree-Ring Research* 59: 21–27.

Meinecke, Emilio. 1926. "Memorandum on the Effects of Tourist Traffic on Plant Life, Particularly Big Trees, Sequoia National Park, California." May 13–16. Unpublished report, USDA–Forest Service Library, Region 5, Vallejo, CA.

———. 1927. Letter to Stephen T. Mather, September 30. Reprinted in Hartesveldt 1962, 298–302.

Muir, John. 1876. "God's First Temples: How Shall We Preserve Our Forests?" *Sacramento Daily Union*, February 5. Reprinted in Grunsky 1968, 242–5. [See also Kimes and Kimes, number 55.]

———. 1877. "On the Post-Glacial History of Sequoia Gigantea," *Proceedings of the American Association for the Advancement of Science* 24: 242–53. [See also Kimes and Kimes, number 63.] [Also issued in offprint form with page numbers running from 3 to 15; the footnote pagination references here refer to the offprint edition.]

———. 1878. "The New Sequoia Forests of California," *Harper's New Monthly Magazine* 57 (342): 813–27. [See also Kimes and Kimes, number 80.]

———. 1894. *The Mountains of California*. New York: Century Company. [See also Kimes and Kimes, number 189.]

———. 1901. *Our National Parks*. Boston: Houghton Mifflin. [See also Kimes and Kimes, number 237.]

———. 1912. *The Yosemite*. New York: Century Company. [See also Kimes and Kimes, number 308.]

Naef, Weston. 2008. *Carleton Watkins in Yosemite*. Los Angeles: J. Paul Getty Museum.

National Geographic Magazine. 1916. Photograph. "The Oldest Living Thing." Vol. 29, no. 4, image facing page 327.

National Park Service. 2004. *Cultural Landscapes Inventory, Mariposa Grove, Yosemite National Park*. Oakland, CA: National Park Service.

O'Connell, Jay. 1999. *Cooperative Dreams: A History of the Kaweah Colony*. Van Nuys, CA: Raven River Press.

Ogden, Kate Nearpass. 1993. "Sublime Vistas and Scenic Backdrops: Nineteenth-Century Painters and Photographers at Yosemite," in *Yosemite and Sequoia: A Century of California National Parks*, 49–68. Berkeley and San Francisco: University of California Press and the California Historical Society.

Olmsted, Frederick Law. 2009. *Yosemite and the Mariposa Grove: A Preliminary Report, 1865*. With an introduction by Victoria Post Ranney. Yosemite National Park: Yosemite Association.

Orsi, Richard. 2005. *Sunset Limited: The Southern Pacific Railroad and the Development of the American West*. Berkeley: University of California Press.

Otter, Floyd. 1963. *The Men of Mammoth Forest: A Hundred-Year History of a Sequoia Forest and Its People in Tulare County, California*. N.p.: self-published.

Otter, Floyd, and David Dulitz. 2007. *The History of a Giant Sequoia Forest: Mountain Home Demonstration State Forest*. Springville, CA: Jason Otter.

Pinchot, Gifford. 1899. *A Primer of Forestry: Part I, The Forests*. Washington, DC: US Department of Agriculture.

———. 1907. *The Use of the National Forests*. Washington, DC: US Department of Agriculture, Forest Service.

Piirto, Douglas D., and Robert R. Rogers. 1999. *An Ecological Foundation for Management of National Forest Giant Sequoia Ecosystems*. Berkeley: US Forest Service, Pacific Southwest Research Station.

Powell, Lawrence Clark. 1989. *California Classics: The Creative Literature of the Golden State*. Santa Barbara: Capra Press.

Runte, Alfred. 1990. *Yosemite: The Embattled Wilderness*. Lincoln: University of Nebraska Press.

Santa Fe Railway, Passenger Department. 1901. *The San Joaquin Valley of California in 1901*. Los Angeles: Kingsley-Barnes and Nuener Company.

Sargent, Shirley. 1964. *Galen Clark: Yosemite Guardian*. San Francisco: Sierra Club.

———. 1975. *Yosemite and Its Innkeepers: The Story of a Great Park and Its Chief Concessionaires*. Yosemite National Park: Flying Spur Press.

———. 1979. *Yosemite's Historic Wawona*. Yosemite National Park: Flying Spur Press.

Sellars, Richard West. 1997. *Preserving Nature in the National Parks: A History*. New Haven, CT: Yale University Press.

Sherwood, George H. 1927. *The Big Tree and Its Story*. New York: American Museum of Natural History.

Southern Pacific Company, Passenger Department. 1914. *Big Trees of California*. San Francisco: Southern Pacific Company.

Stephenson, Nathan L. 1996. "Ecology and Management of Giant Sequoia Groves," *Sierra Nevada Ecosystem Project, Status of the Sierra Nevada* 2: 1431–67.

Stewart, George W. 1930. *Big Trees of the Giant Forest: Their Life Story from the Blossom Onward*. San Francisco: A. M. Robertson.

Stine, Scott. 2015. *A Way Across the Mountains: Joseph Walker's 1833 Trans-Sierra Passage and the Myth of Yosemite's Discovery*. Norman, OK: Arthur H. Clark Company, an imprint of the University of Oklahoma Press.

Strong, Douglas. 1964. "A History of Sequoia National Park." PhD diss., Syracuse University.

Sudworth, George B. 1908. *Forest Trees of the Pacific Slope*. Washington, DC: US Department of Agriculture, US Forest Service.

Tweed, William C. 1972. "Sequoia National Park Concessions: 1898–1926," *Pacific Historian* 16 (1): 36–60.

———. 1992. "Public Perceptions of Giant Sequoia Over Time," *Proceedings of the Symposium on Giant Sequoias: Their Place in the Ecosystem and Society*, 5–7. Berkeley: US Forest Service, Pacific Southwest Research Station.

———. 2010. *Uncertain Path: A Search for the Future of National Parks*. Berkeley: University of California Press.

US Congress. House of Representatives. Committee on the Public Lands. 1939. *Establishing the John Muir–Kings Canyon National Park, California: Hearings Before the Committee on the Public Lands, House of Representatives*. Washington, DC: GPO.

US Department of Agriculture, Division of Forestry. 1900. *A Short Account of the Big Trees of California*. Washington, DC: GPO.

US Department of the Interior. 1915. *The Sequoia and General Grant National Parks, Season of 1915*. Washington, DC: GPO.

———. 1921. *Rules and Regulations, Sequoia and General Grant National Parks, 1921*. Washington, DC: GPO.

———. 1924. *Rules and Regulations, Sequoia and General Grant National Parks, 1924*. Washington, DC: GPO.

US Forest Service. 1991. *Sequoia National Forest, Giant Sequoia Management: Before the Subcommittee on General Oversight and California Desert Lands, Committee on Interior and Insular Affairs, United States House of Representatives*. Berkeley: US Forest Service, Pacific Southwest Research Station.

———. 1994. *Proceedings of the Symposium on Giant Sequoias: Their Place in the Ecosystem and Society*. Berkeley: US Forest Service, Pacific Southwest Research Station.

———. 2012. News release. "Giant Sequoia National Monument Management Plan Promotes Protection and Ecological Restoration through Science and Collaboration." September 4.

US House of Representatives. 1900. *Letter from the Secretary of the Interior, with a Draft of a Proposed Bill, A Statement of Facts Relating to Negotiations for Purchase of Certain Groves of Sequoia Gigantea in California*. Washington, DC, 56th Cong., Doc. No. 626.

Van Name, Willard. 1927. *The Redwood Mountain Sequoia Grove: The Third Largest Grove of Big Sequoias in the World, Excelled Only by the Garfield and Giant Forest Groves.* N.p.: Privately printed.

Vermaas, Lori. 2003. *Sequoia: The Heralded Tree in Art and Culture.* Washington, DC: Smithsonian Books.

Vischer, Edward. 1862. *Vischer's Views of California: The Mammoth Tree Grove and Its Avenues.* San Francisco: self-published.

Weiner, Jackie. 2009. *Timely Exposures: The Life and Images of C. C. Curtis, Pioneer California Photographer.* N.p.: self-published.

Wells, Andrew Jackson. 1907. *Kings and Kern Canyons and the Giant Forest of California.* San Francisco: Southern Pacific Company.

White House. 2000. News release. "Remarks by the President at Ropeline Following Earth Day Event." April 15.

Whitney, Josiah. 1869. *The Yosemite Book: A Description of the Yosemite Valley and the Adjacent Regions of the Sierra Nevada and of the Big Trees of California.* New York: Julius Bien. [Online at http://www.yosemite.ca.us/library/the_yosemite_book.]

Willard, Dwight. 1995. *Giant Sequoia Groves of the Sierra Nevada: A Reference Guide.* Second edition. Berkeley: self-published.

———. 2000. *A Guide to the Sequoia Groves of California.* Yosemite National Park: Yosemite Association.

Willis, Nathaniel Parker. 1859. "The Mammoth Trees of California," *Hutchings' Illustrated California Magazine* 3 (March): 385–97. [Online at http://www.yosemite.ca.us/library/hutchings_california_magazine/33.pdf.]

Wilson, Herbert Earl. 1922. *The Lore and the Lure of the Yosemite.* San Francisco: Sunset Press.

———. 1928. *The Lore and the Lure of Sequoia: The Sequoia Gigantea, Its History and Description.* Los Angeles: Wolfer Printing Company.

Wolfe, Linnie Marsh. 1938. *John of the Mountains: The Unpublished Journals of John Muir.* Boston: Houghton Mifflin. [See also Kimes and Kimes, number 379.]

Wood, Harold. 2000. "President Clinton Proclaims Giant Sequoia National Monument," *Road Runner* 48 (May).

Worster, Donald. 2008. *A Passion for Nature: The Life of John Muir.* Oxford, UK: Oxford University Press.

Index

Act to Set Apart Certain Tracts of Land in the State of California as Forest Reservations, An (national park bill of 1890), 97–99
Adams, Alvin, 11
advertising and marketing, 13–14, 15–18, 26, 92–93, 101, 124–26
age of the trees, 16, 66, 125, 126–27, 129, 130, 140–41, 144, 169, 170, 172, 173–74, 176, 184; see also dendrochronology
Albright, Horace, 135, 151, 157, 158, 159–60, 162, 163, 196, 216, 218
Alden, James Madison, 34
Allen, B. F., 113–14
American Academy of Arts and Sciences, 24
American Association for the Advancement of Science, 63, 64, 68–70, 128
American Journal of Science and Arts, 24
American Museum of Natural History, 102, 242n23
American Naturalist, 63
Antiquities Act, 225, 226
Arbor Vitæ, 5, 26
art of the trees, 46–48, 132–34
Atchison, Topeka, and Santa Fe Railway, 125–26
Audubon, 221
Audubon Society, 222
automobile tourism, 148–50, 153, 166, 215–17; see also tourism
Ayres, Thomas A., 34–35

Backcountry Horsemen of America, 222
Baker, Lucretia, 52
Balch, Allan and Janet, 185–86
Balch Park, 186
bald cypress, 63
Ballinger, Richard, 134
Bearskin Meadow Grove, 219
Bellue, Alfred J., 157
Bierstadt, Albert, 46–47
Big Oak Flat Road, 93
Big Stump, 1–2, 14, 16, 17; see also Discovery Tree
Big Stump Basin and Grove, 76–77, 79, 82, 96, 102, 189, 242n23
Big Tree Cottage, 14, 16

Big Tree mania, 26
Big Trees, 85, 174–179, 203
Big Trees Lodge, 150–51, *152*, 160–61, *161*, 162, 188, 216
Big Trees of California, The, 125, 126–128
Big Trees of the Giant Forest, 173–74
Big Trees Station, 42, 44, 48; see also Wawona Hotel
Biswell, Harold, 208–9
Black Mountain Grove, 196, 246n35
booklets and guidebooks, 9, 17, 125, 165–67, 169, 238n10; see also advertising and marketing
Boole Tree, 190
Boston Evening Transcript, 36
botanical interest in and scientific details about the trees, 22–27, 60, 61–63, 65–67, 68–71, 126–27
bowling on a felled tree, 14, 16
British Museum, 102
Britton and Rey, 5
Brower, David, 210
Brown, Edmund G. "Pat," Sr., 197
Brownie the mule, 64, 66, 68, 70, 130
Bush, George H. W., 223–24
Bush, George W., 226

Cain, Stanley, 204
Calaveras Big Trees, 2, 3–5, 15–19, *18*, 22–23, 30, 33, 35, 39, 46, 48, 62, 69, 94–95, 115–17, 120, 125, 186–87, 194–96
Calaveras Bigtrees National Forest (proposed), 186
Calaveras Grove Association, 186–87
Calaveras National Park (proposed), 115–18
California Academy of Sciences (also California Academy of Natural Sciences), 6, 22, 236n1
California Native Plant Society, 222
California redwood, see coast redwood
California Tunnel Tree, 241n4
Cammerer, A. B., 158, 162
Camp Kaweah, 159
Camp Sequoia, 108
Camp Six, 140, 144
Carr, Ezra, 62

259

Carr, Jeanne, 62, 239n6
Centennial International Exhibition, 47, 55–56, 66
Centennial Stump, 56, 144; see also Daniel Webster Tree
Century Magazine, 97, 112
Chase, J. Smeaton, 131–32
Chicago Stump, 242n23
Civil War, 18, 38, 46
Clark, Galen, 30–33, *31*, 34, 43–44, 46, 60, 94, 126–27, 131, 140, 172, 188
Clark's "hospice," 44, 151, 161, 205
Clark's Station, 32, 33, 43–44
climate and climate change, 63–64, 139–45, 168, 176, 230–31
Clinton, Bill, 224–26
coast redwood, 8–9, 25, 63, 197
Comstock, Smith, 76, 82, 102
Conness, John, 36, 39
conservation, preservation, and restoration, 13, 35, 36–39, 53–57, 71–72, 78, 89, 95–97, 100, 101, 108–10, 112–15, 116–18, 119–21, 134, 154–62, 182–97, 203–8, 215–18, 220–28
Converse Basin and Grove, 66, 71, 80, 83–86, 96, 103–4, 114, 116, 120, 138, 182, 192, 196, 220
Cook, Lawrence, 157
Co-operative Land Purchase and Colonization Association, 87–88; see also Kaweah Colony
Coulterville Road, 93
Crane Flat, 3
Crater Lake National Park, 134
Crystal Palace (London), 11, *12*, 39, 236n24
Crystal Palace (New York), 7, 11
Curtis, C. C., 102, *103*, 104, *104*, 242n22
cypress, 63

dancing on a stump, 7, 14, 16
Daniel Webster Grove, 56–57, 66, 244n15 (chap. 11)
Darwin, Charles, 62
Dead Giant, 51, 54, 93–94, *93*, 241n3
Deer Creek Grove, 68
dendrochronology, 138–44
Depression, 173, 187–88, 189
Descaisne, Joseph, 25
Desmond, D. J., 150
Desmond Company, 150
Dillonwood Grove, 79, 81, 120, 195, 196
Dinkey Creek, 239n19
Dinky Grove, 239n19
Discovery Tree, 2, 4, 5–9, 13, 14, 16, 17, 26, 235n10, 237n14
Douglass, Andrew Ellicott, 138–40, 142–44, *143*, 169, 170

Dowd, Augustus T., 3–4, 5, 15, 22, 30n
Drew, Arthur H., 192–93
Drury, Newton, 196
Dudley, William R., 117
Duke of Wellington, 24

El Portal, 124, *124*
Ellsworth, Rodney Sydes, 170–71, 202
Emerson, Ralph Waldo, 43
Endlicher, Stephen, 25, 170–71, 239n11 (chap. 6)
Enterprise Mill, 140, 144
Evans Grove, 138, 241n21
evolution, 62–63, 69–71
exhibits of trees, 6–9, *10*, 11, *12*, 13–14, 39, 54–56, 102–105, *103*, *104*, *105*

Fallen Wawona Tunnel Tree, 92, 241n3; see also Wawona Tunnel Tree
Father of the Forest, 17
Federal Aid Road Act, 149
Ferguson, John W., 55
fire, 72, 101, 110–12, 129, 130, 169, 172, 177, 200–3, 206–11, 215, 219, 223
First Cavalry, 108
Fisher, Walter, 134
Forest and Related Resources (FARR) Plan, 247n10
Forest King, 54–55
Forest Queen, see Galen Clark Tree
Forest Service, creation of, 114
Forest Trees of the Pacific Slope, 201
Freeman Creek Grove, 196, 223–24, 246n35
Fresno Grove, 33, 53, 65, 69, 71, 78–79; see also Nelder Grove
Fresno Junior Chamber of Commerce, 190
Fricot, Désiré, 186–87
Fry, Walter, 85, 174–79, *175*, 203

Gale, George, 9, 11, 13
Galen Clark Tree, 29, 33
Gardeners' Chronicle, 23
Garfield Grove, 96–97
Gearhart, Bud, 191
General Grant National Park, 99, 100, 101, 102–3, *109*, 114, 115, 125, 153, 167, 191
General Grant Tree, 52, 56–57, 100, *109*, 144, 182, 190
General Land Office, 80, 81, 96, 100, 112, 113; Visalia district, 54–55, 81, 87–88, 95–96
General Noble Tree, 103–5, *103*, *104*, *105*, 242n23
General Sherman Tree, xv–xvi, xvii, 172, 173–74, 182, 184, 190
Genesis Tree, 182
George Bush Tree, 223

Giant Forest, 53, 67, 80, 82, 86–88, 100, 107–8, 125, 130–31, 134, 135, 151, 153–54, 155, 158–59, 163, 165, 173–74, 175, 183, 184, 216–18, 246n23
Giant Forest Lodge, 147–48, 151, 152, 159–60, *159*, 161, 218
Giant Redwood Trees of California, 47
Giant Sequoia, The, 170, 202
Giant Sequoia National Monument, 214–15, 225–28
Giant Sequoia of the Sierra Nevada, The, 210
Giant Tree exhibit, 6; see also Discovery Tree
Gillett, Frederick, 135
Glacier National Park, 134
Glacier Point Hotel, 150
Glyptostrobus pensilis, 63
gold rush, 2, 3, 4, 30, 52, 77–78
Gore, Al, 226
Grant Grove, 54, 66, 79, 95–96, 98, 120, 238n2 (chap. 5), 240n22; see also Tulare Grove
Graves, Henry, 185
Gray, Asa, 22, 24, 62–64, 68, 70, 71
grazing animals, 109–10; see also sheep
Great Depression, 173, 187–88, 189
Great Trees, Mariposa Grove, California, The, 47, 48n
Greeley, Horace, 35
Grinnell, Joseph, 169
Grizzly Giant Tree, 34, 44, 46, 47, 92, *119*, 127, 131, 133, 155–57, *156*, 215
Grosvenor, Gilbert, 135, 183–84
grove (definition of), 2n, 178
guidebooks, see booklets and guidebooks
Guide to the Giant Forest, Sequoia National Park, 168
Guide to Yosemite, 168
Gwin, William, 11

Hall, Ansel, 168–69
Hammond, Natalie Harris, 185
Handbook of Yosemite National Park, 168, 169, 202
Hanford, William H., 5–9, 23
Harper's New Monthly Magazine, 60, 70, 128, 129
Harriman, Edward, 119
Harrison, Benjamin, 97, 113
Hart Tree, 190
Hartesveldt, Richard J., 204–11, 215
Harvey, H. Thomas, 208, 210
Haynes, A. Smith, 14
Hays, Howard, 159–60, 161–62
Hercules Tree, 17
Hetch Hetchy Valley, 134
High Sierra Camp, 244n3 (chap. 12)

Hill, Estella, 133
Hill, Thomas, 132–33
Hoist Ridge, 241n16
Hubbard, Douglass, 204
Hume, George, 185, 188–89, 192
Hume, Thomas, 138, 182, 192
Hume-Bennett Lumber Company, 137–38, 140–41, 182, 189, 192, 245n5
Hume Lake, 85, 138, 182, 185, 188–89, 192
Huntington, Collis P., 102, 169, 170
Huntington, Ellsworth, 139–44, 167–68
Hutchings, James Mason, 15, 16, 34, 45, 61, 117
Hutchings' Illustrated California Magazine, 15, 34
Hyde's Mill, 67, 79, 81, 189, 246n27 (chap. 15)

Ickes, Harold, 191
Indians of Yosemite Valley and Vicinity, 126
Iowa Tree, 203

Jepson, Willis Linn, 169, 170, 178, 202
Jewett, William Smith, 34
Johnson, Robert Underwood, 97
Jordan, David Starr, 116
Jourdan, J. W., 190

Kaweah Colony, 88–89, 100–1, 183; see also Co-operative Land Purchase and Colonization Association
Kaweah River watershed, 52–53, 67–68, 79, 80, 82, 86–88, 96, 98, 113, 130, 184
Kellogg, Albert, 22–24
Kern Canyon, 125
Kern River watershed, 214, 242n7 (chap. 8)
Kilgore, Bruce, 209–10
King, Thomas Starr, 36
Kings Canyon National Park, 51–52, 76, 125, 191, 199–200, 208, 209
Kings River Canyon, 82, 137–38, 185
Kings River Grove, 66, 71
Kings River groves, 79, 82, 83, 120
Kings River Lumber Company, 83–85
Kings River Parks Company, 151
Kings River watershed, 79, 113, 120, 189, 238n11
Kroeber, A. L., 169

Lamson, James, 34
Lane, Franklin, 134–35
Lapham, Joseph, 4–5, 26
Lapham, William, 4–5, 9, 14
Lawson, A. C., 169
Leonard, Zenas, 3
Leopold, A. Starker, 209–10
Leopold Report, 209–10
Leslie, Frank, 54

Lindley, John, 23–25
Lindsey, Tipton, 95
literature of the trees, 124–32, 166–79; see also booklets and guidebooks, and entries listed by author or publication
Lobb, William, 6, 22–24, 237n11
Lockyer, Bill, 227
logging, 67, 71–72, 75–89, 95–97, 101, 129, 181, 192–94, 195–96, 214–15, 219–28
Long Meadow Grove, 214, 225
Lore and the Lure of Sequoia, The, 172
Lore and the Lure of the Yosemite, The, 171–72
Low, Frederick, 37, 42
Lowell, Percival, 139
lumber industry, see logging

Mammoth Grove Hotel, 14, 15, *18*, 115
Mammoth Tree exhibit, 6; see also Discovery Tree
management of the parks, see park management
Mann, Houston, 32
Mann, Milton, 32
Mariposa Battalion, 237n3, 237n8
Mariposa Grove, 30, 33, 34, 35–39, 42, 43–46, 48, 60, 62, 65, 69, 92, 94, 94, 95, 99, 111, 114, 115, 117, 116, 118, 119–20, *119*, 130, 150–51, *152*, 155, 157, 160, *161*, 163, 170, 203–4, 205–7, 215–16
Mariposa Grove Museum, 30
Mark Twain Tree, 102, 203, 242n23
marketing, see advertising and marketing
Massachusetts Tree, 203
Mather, Stephen T., 134–36, 148, 149–50, 151, 153, 155, 158–59, 162, 163, 166, 168, 169, 183, 184–85, 186, 216, 218
Mather Mountain Party, 135
Matthew, John D., 26
McCormick, Ernest O., 135
McKinley Grove, 239n19
Meinecke, Emilio, 152–55, 157–58, 160, 161, 163, 203, 205, 216
Meinecke report, see Meinecke, Emilio
Merced Grove, 93, 99, 111, 114
Merced River, 31–32, 43–44, 112
Mesa Verde National Park, 134
Michigan Trust Company, 192–93
Mill Flat, 52, 54, 66, 79, 83
Miller, Adolph, 134–35
Miller, John, 241n7
Mills, William H., 111
Miwok, 2, 111, 237n8 (chap. 3)
Montana Tree, 151, *152*
Moore, Austin, 82–83, 103
Moore, Edwin, 43
Moore Cottage, 42
Mother of the Forest, 9–13, *10*, *12*, 14, 16, 17, 39, 55, 237n14
Mountain Home Demonstration State Forest, 182, 193, 195
Mountain Home Grove, 79, 81, 120, 140, 144, 185, 192–94
Mountains of California, The, 128–29, 131, 200
Mt. Rainier National Park, 134
Muir, John, 56, 59–62, 64–72, 78, 80–81, 82, 97, 98, 109–10, 111, 112–13, 116, 117, 118–19, *119*, 128–31, 134, 140, 168–69, 172, 200–1
Muir Snag, 59–60, 66, 239n21
Murphy, John, 3
Murphys, 3, 4, 15
Museum of Natural History, 242n23

naming of the Big Trees, 22, 23–26; see also *Sequoia* name
National Geographic, 183–84, 195, 221
National Geographic Society, 183–84, 195
National Mall, 105, 242n23
national park bill of 1890, 97–99
National Park Service, creation of, 135–36, 149; uniforms, 136
national park system, creation of, 37–38, 95, 97–101, 108, 114–15, 120, 134, 241n5
National Parks, 208, 210
National Science Foundation, 205
National Wildlife Federation, 205
Native Americans, 2, 5, 33, 47, 53, 111, 171–72, 200
Natural Resources Defense Council, 227
Nelder, John A., 65, 70
Nelder Grove, 54n, 65, 78–79, 237nn8–9 (chap. 3); see also Fresno Grove
Nelson, DeWitt, 193
"New Sequoia Forests of California, The," 68, 70–71, 128
New Zealand, 21, 26
Niles Canyon Railroad, 124
Nixon, Richard, 247n10
Noble, John W., 99, 100, 103, 108, 113, 114, 116
North Calaveras Grove, see Calaveras Big Trees

Observation Car 330, 123–24
Ogg, R. H., 32
Old Adam snag, 82
Olmsted, Frederick Law, 37–38, 43
"On the Post-Glacial History of Sequoia Gigantea," 68–69, 240n27
Osborn, Henry Fairfield, 135
Our National Parks, 81, 128, 129–31, 167, 239n17

Overland Journey from New York to San Francisco in the Summer of 1859, An, 35
Overland Monthly, 64

park management, 108–12, 118, 134–35, 148–52, 157–62, 200–5, 207, 209–10, 215–28
People's Mill, 79
Perkins, George, 116
Perry, John, 14
Pickering Lumber Company, 186, 194, 196
Pillsbury, E. S., 116
Pinchot, Gifford, 114, 201–2, 219
Pioneer Cabin Tree, 95
preservation, see conservation, preservation, and restoration
Preston, John, 204, 205
Primer of Forestry, A, 202

Queenstown, 21, 26

railroad, 18–19, 78–79, 83, 123–26; see also Southern Pacific Railroad
Raymond, 99, 115, 118, 133
Raymond, Israel Ward, 36–38, 43
redwood, see coast redwood
Redwood Canyon, 189–91, 208
Redwood Mountain, 67, 79, 200
Redwood Mountain Grove, 189–91, 199, 208, 246n27
restoration, see conservation, preservation, and restoration
Rockefeller, John D., Jr., 187, 196
Roosevelt, Franklin, 191, 199
Roosevelt, Theodore, 114, 118–20, *119*, 133, 185
Round Meadow, 159–60, 245n3

San Francisco Tree, 203
San Joaquin Light and Power Company, 185
Sanger Lumber Company, 84–85, *84*, 103–4, 138, 185
Save the Redwoods League, 187
Scenes of Wonder and Curiosity in California, 45
Scribner's Magazine, 102
Secret of the Big Trees, The, 167–68
sequence tree, 25
Sequoia name, 25, 127, 170–71, 172
Sequoia and Kings Canyon National Parks, 210, 217
Sequoia Creek Grove, 79
Sequoia gigantea, 25
Sequoia Lake, 239n20
Sequoia National Forest, 182, 189, 192, 196, 214–15, 219–21, 222, 223–28
Sequoia National Park, 99, 100, 101, 107–8, 110, 112, 114, 125, 135, 147, 151–55, *156*, 158–59, *159*, 165, 168, 172, 173, 174–78, 183–85, 188, 190, 216–18
Sequoia Railroad, 83
Sequoia sempervirens, 25
Sequoia washingtoniana, 237n9
sequoia wine, 61, 239n6
Sequoiadendron giganteum, 21, 237n9
Sequoyah (Cherokee Indian), 127, 170–71, 172
sheep, 70, 72, 101, 109–10
Shellhammer, Howard, 208
Sheridan Tree, 155
Sierra Club, 116, 117, 118, 134, 189, 210, 221, 222, 227
Sierra Club Bulletin, 59–60, 208, 210
Sierra Forest Reserve, 113–14, 115, 120
Sierra Nevada Ecosystem Project, 224
Smith Comstock Mill, 82
Smith, Hiram, 82–83, 103
Smithsonian Institution, *105*, 242n23
Snediker, William, 53–56
Société Botanique de France, 25
Soldiers Camp, 107–8
Soldiers Trail, 107–8
South Calaveras Grove, see Calaveras Big Trees
Southern Pacific Railroad, 98–99, 102, 112, 115, 119, 125, 135, 190
Sparks, William, 87
Spencer, Eldridge, 161
Sperry, J. C., 187
Sperry, James L., 14, 115–16
Sperry's Hotel, 15, 18, 186
Stable Tree, 203
staircase built onto a tree, 16, 17
state park system, 187, 196
state tree status, 197
Stecker, Ronald E., 208
Stegman, William, 53–56
Stewart, George, 96, 98, 100, 112–13, 173–74
Stewart, Ronald, 223
Strenzel, Louisa, 239n6
"Studies in the Sierra," 64
Sudworth, George B., 117, 189, 201, 237n9
"Summering in the Sierra," 64
Sumner, Lowell, 204
Sunset, 221
Swamp and Overflow Act, 183
"Symposium on Giant Sequoias, The," 223

Taft, William Henry, 123, 202
Taxodium distichum, 63
Taxodium sempervirens, 25
Texas Tree, 155
Tharp, Hale, 53, 130
Thomas's Mill, 66, 83

Thompson, Charles G., 155, 157–58, 160–61
Timber and Stone Act, 80–83, 87–89, 95–96, 100–1, 113–14, 120, 138, 182, 183, 189
Timber Culture Act, 113
timber industry, see logging
Tirrell, George, 34
Torrey, John, 22, 62
tourism, 14–17, 18, 18–19, 32–33, 36, 43, 44–45, 46, 48, 78, 91–95, 98, 99–100, 101, 109, 115, 117, 123–26, 148–50, 153–62, 156, 166–75, 178, 179, 203–7, 215–18; see also automobile tourism
Trail of a Hundred Giants, 214, 225
Trask, George, 9, 11, 13
tree rings, see dendrochronology
Trees of California, The, 178
Tryon, John, 8
Tulare County groves, 78–79, 95–96, 98, 116, 173
Tulare Grove, 52–57, 240n22; see also Grant Grove
Tule River groves, 71, 80–81, 82
Tule River watershed, 53, 68, 79, 80–81, 113, 120, 191–93, 195
tunnel trees, see Dead Giant, Pioneer Cabin Tree, and Wawona Tunnel Tree
Tuolumne Grove, 93, 99, 111, 114
Tuolumne Meadows, 150
Tuolumne River watershed, 112

Union Water Company, 4
United States Forest Service, creation of, 114
US Army, 107–12
Utah Tree, 203
utopian society, see Kaweah Colony

Vacation among the Sierras, A, 237n15
Van Name, Willard, 190–91
Vandever, William, 96, 97–98, 99
Veitch & Sons, 22, 23, 26
Visalia General Land Office, see General Land Office, Visalia district
Vischer, Edward, 17, 18
Vischer's Views of California, 17
Vivan, Martin, 55–56, 244n15 (chap. 11)

Wagner, Theodore, 95
Walker, Frank J., 95
Walker, Joseph, 3, 241n1
Walton, Strother, 192
Warren, Earl, 193
Washburn, Estella Hill, 133
Washburn family, 133, 150, 187–88
Washburn, Henry, 44, 94, 95
Washburn, Jean Bruce, 44, 45
Washburn, John, 133

Washington cedar, 24
Washington, DC, 105, 242n23
washingtoniana, 237n9
Watkins, Carleton, 34–35, 34, 46, 237n11 (chap. 3)
Wawona Hotel, 41–42, 44–45, 48, 94, 115, 118, 133, 150, 187–88
Wawona Tunnel Tree, 91–92, 94, 115, 133, 203, 215, 216, 241n3
Weed, Charles, 34–35
Wellesley, Arthur, 23
Wellingtonia gigantea, 21–22, 24, 26
Western Mono, 2, 111
Wheeler, [no first name], 35
Whitaker, Horace, 240n24
Whitaker's Forest, 209, 240n24
White, John R., 85, 153, 155, 157, 158–60, 163, 174–79, 175, 203, 217
Whiteside, Robert, 116, 117, 186–87
Whitney, Josiah, 43, 45–46, 53, 117
Wilderness Society, 222
"Wildlife Management in the National Parks," see Leopold Report
Willett, Edward W., 54
Willis, Nathaniel P., 15–16, 18
Wilson, Herbert Earl, 171–73
Wilson, Woodrow, 134, 136, 149
Wolverton, James, 130
World's Columbian Exposition, 103–5, 103, 133
Wright, George M., 158

Yellowstone National Park, 114, 116, 241n5
Yosemite, The, 110, 128, 131
Yosemite bill of 1890, 97–98
Yosemite Book, The, 45–46, 53
Yosemite Park Commission, 43, 112, 118, 126
Yosemite Grant, 36–39, 42, 45, 96, 97–98, 99, 111, 112, 118
Yosemite Lodge, 150
Yosemite Museum, 166
Yosemite National Park, 99–100, 101, 109–10, 111, 112, 114, 115, 118–19, 124, 134–35, 149–50, 155, 160, 162, 168, 169, 171–72, 187–88, 203, 210, 215
Yosemite National Park Company, 150, 151, 188
Yosemite Park Commission, 111, 112
Yosemite Trails, 131–32
Yosemite Valley, 31–39, 43, 44, 45, 47, 60–61, 62, 64, 93–94, 97, 99, 101, 111, 115, 118, 119, 125, 127, 132–33, 168
Yosemite Valley Railroad, 123–24

Zumwalt, Daniel K., 98–99

About the Author

William C. Tweed brings humans closer to nature using the knowledge and skills he developed during thirty years as a chief naturalist, historian, and writer with the National Park Service. His published works include *Sequoia-Kings Canyon National Parks: The Story Behind the Scenery* (KC Publications, 1980); *Challenge of the Big Trees: A Resource History of Sequoia and Kings Canyon National Parks* (coauthored with Lary Dilsaver; Sequoia Natural History Association, 1990); *Death Valley and the Northern Mojave: A Visitor's Guide* (coauthored with Lauren Davis; Cachuma Press, 2003); and *Uncertain Path: A Search for the Future of National Parks* (University of California Press, 2010). Tweed makes his home in Bend, Oregon.

About the Sierra College Press

In 2002, the Sierra College Press was formed to publish *Standing Guard: Telling Our Stories* as part of the Standing Guard Project's examination of Japanese American internment during World War II. Since then the Sierra College Press has grown into the nation's first academic press operated by a community college.

The mission of the Sierra College Press is to inform and inspire scholars, students, and general readers by disseminating ideas, knowledge, and academic scholarship of value concerning the Sierra Nevada region. The Sierra College Press endeavors to reach beyond the library, laboratory, and classroom to promote and examine this unique geography.

For more information, please visit www.sierracollege.edu/press.

Editor-in-Chief: Joe Medeiros

Board of Directors: Chris Benn, Rebecca Bocchicchio, Keely Carroll, Kerrie Cassidy, Mandy Davies, Dan DeFoe, David Dickson, Dave Ferrari, Tom Fillebrown, Rebecca Gregg, Brian Haley, Robert Hanna, Rick Heide, Jay Hester, David Kuchera, Joe Medeiros, Lynn Medeiros, Sue Michaels, Gary Noy, Bart O'Brien, Sabrina Pape (Chair), Mike Price, Jennifer Skillen, Barbara Vineyard.

Advisory Board: Terry Beers, Frank DeCourten, Patrick Ettinger, Tom Killion, Tom Knudson, Gary Kurutz, Scott Lankford, John Muir Laws, Beverly Lewis, Malcolm Margolin, Mark McLaughlin, Bruce Pierini, Kim Stanley Robinson, jesikah maria ross, Mike Sanford, Lee Stetson, Catherine Stifter, Rene Yung.

Special thanks to Sierra College Friends of the Library, a major financial supporter.